Natural Fiber

Edited by Han-Yong Jeon

Published in London, United Kingdom

Natural Fiber
http://dx.doi.org/10.5772/intechopen.97897
Edited by Han-Yong Jeon

Contributors
Yamini Jhanji Dhir, Tahlaiti Mahfoud, Nouri Mustapha, Ramratan Guru, Anupam Kumar, Rohit Kumar, Richard Ntenga, Martin Tango, Etienne Mfoumou, Pagore Djoda, Saidjo Saidjo, Annie Wakata, Palash Mallick, Susanta Sekhar De, Kavita Mathur, Ruksana Baby, Emiel DenHartog, Malgorzata Zimniewska, Barbara Romanowska, Daniel Levacher, Mazhar Hussain, Huyen Bui

First published in London, United Kingdom, 2022 by IntechOpen
IntechOpen is the global imprint of INTECHOPEN LIMITED, registered in England and Wales, registration number: 11086078, 5 Princes Gate Court, London, SW7 2QJ, United Kingdom

British Library Cataloguing-in-Publication Data
A catalogue record for this book is available from the British Library

Additional hard and PDF copies can be obtained from orders@intechopen.com

Natural Fiber
Edited by Han-Yong Jeon
p. cm.
Print ISBN 978-1-80355-213-2
Online ISBN 978-1-80355-214-9
eBook (PDF) ISBN 978-1-80355-215-6

For EU product safety concerns:
IN TECH d.o.o., Prolaz Marije Krucifikse Kozulić 3, 51000 Rijeka, Croatia,
info@intechopen.com or visit our website at intechopen.com.

Meet the editor

Professor Emeritus Han-Yong Jeon is a geosynthetics/technical organic materials researcher at Inha University, South Korea. He has been the director of the Geosynthetics Institute (GSI), USA, since 1998. Previously, he was a council member for the International Geosynthetics Society and president of the Korean Geosynthetics Society as well as the Korean Fiber Society. He has published more than 1,000 proceedings in domestic and international conferences and 226 papers in scientific journals. He has also written twenty-five textbook chapters.

Contents

Preface

Compared with synthetic fibers, natural fibers are sustainable, carbon neutral, and environmentally friendly. They have a variety of applications and can be developed into hybrid products that can be used as biomaterials and fusion materials via coating or modification. They can also be recycled and reused.

Research on eco-friendly fibers is being conducted not only in the green technology fields such as organic fibers, renewable fibers, natural fibers, and biodegradable fibers but also in the nanofiber technology fields such as hydrogen battery separators and medical fibers. The potential of natural fibers as building materials for civil engineering is also attracting research attention.

In the past, artificial fibers were usually made of viscose rayon, a regenerated fiber made by dissolving wood pulp, a natural material, in an alkali solution and chemically spinning it. It was the first synthetic fiber with better hygroscopicity than cotton and a luster like silk. Semi-synthetic fibers called acetate or triacetate are made from natural pulp or cotton linter as raw materials and chemically changed.

Among eco-friendly textile materials, biomass-based textile materials are biodegradable but require composite technology for production. As such, a high-efficiency, eco-friendly process that minimizes the use and emission of polluting or hazardous chemicals is needed.

This book presents a comprehensive overview of natural fibers and examines techniques for modifying them so they can be used for practical applications. It discusses the 2D and 3D structure characteristics of a variety of fibers, including cellulose-based natural fibers such as cotton fibers.

We sincerely thank all the contributing authors as well as Author Service Manager Ms. Paula Gavran at IntechOpen.

Han-Yong Jeon
Department of Chemical Engineering,
Inha University,
Incheon, South Korea

Section 1

Overview of Natural Fibers

Chapter 1

Natural Fibers: The Sustainable Alternatives for Textile and Non-Textile Applications

Yamini Jhanji Dhir

Abstract

The increasing environmental concerns and depletion of petroleum resources have increased the importance of natural fibers and have stimulated researchers and industries to use sustainable fibers instead of conventional synthetic fibers. Besides exceptionally brilliant mechanical and physical properties are also attractive aspects of natural fibers enabling the utilization of natural fibers in myriad of textile and non-textile applications such as clothing, and reinforced composite products in various industries such as automotive, building, and furniture. Natural fiber composites are composite materials comprising of reinforcing fibers derived from renewable and carbon dioxide neutral resources such as wood or plants. NFCs find application in molded articles that demand moderate strength for acceptable performance for various indoor and outdoor applications. A rapid drift from oil-derived polymers and mineral-reinforced materials to sustainable alternatives has fostered automotive and packaging industries to start utilizing natural fiber composites in their designs. Accordingly, natural fiber composites are serving as energy efficient and sustainable alternatives replacing traditional materials such as metals, polymeric resins, and reinforcement fibers. A worldwide clamor for green products and thus upsurge in sustainable alternatives have been witnessed as a result of diminishing petroleum reserves worldwide, exorbitant prices of petroleum, and high disposal costs of petroleum-based composites along with inability of decomposition of some petroleum-based composites. Contrastingly, natural materials outshine the petroleum-based products in being renewable, inexpensive, biodegradable, and eco-friendly.

Keywords: natural fibers, sustainable, environment, textile, bio degradability, composites

1. Introduction

The significance of natural fibers to fulfill the basic human needs of clothing and shelter has been well established since time immemorial. However, with the advent of synthetic fibers in 1900s, the popularity and usage of natural fibers became archaic. Nevertheless, the world wide clamor for sustainable and eco-friendly approaches in textile supply chain and depletion of petroleum resources have stimulated the usage of natural fibers thereby replacing synthetic fibers with sustainable natural fibers.

Natural fibers are grown naturally and do not pose any detrimental environmental impact except for situations where fertilizers, pesticide and other toxic chemicals are extensively utilized to improve their yield. The adverse impacts of synthetic fibers on environment and their fossil fuel based origin fosters industrial establishments, researchers and technologists to explore latest and innovative methods of growth, development, cultivation and usage of natural fibers via eco-friendly mode. Cotton holds lion's share of market as far as textile applications are concerned but the fiber cannot be considered sustainable owing to massive utilization of water, pesticides, fertilizers and toxic chemicals thereby leading to environmental and economic distress. Apart from cotton, some resource efficient fibers that are replacing cotton in varied textile applications include linen, hemp, flax, jute and bamboo.

Linen is gaining interest among textile designers for designing clothing, footwear and handbags. Hemp and jute are rope like fibers and exhibit coarseness compared to linen and generally preferred for apparels and accessories that demand rough texture and durability. Jute is a plant based multicellular fiber characterized by nodes and cross markings in longitudinal view and polygon shapes in cross-section. *Flax* is cellulosic fiber in crystalline form featuring a length of 90 cm and diameter of 12–16 μm. Flax is mainly cultivated in Canada, Netherlands, Belgium and France. The stem of plant *Linum usitatissimum* is the fiber source. The fiber extraction is accomplished by two processes namely retting and scorching to alter the fiber properties. The enzymatic application during retting process causes pectin degradation thereby resulting in separation of fibers. Flax fiber is extensively used for production of linen besides being used in furniture, home textiles and interior decor items.

Bamboo and straw are another plant based fibers gaining popularity among fashion designers. Bamboo is wood like tropical grass while the sources for straw are wheat stalks, grasses, sisal hemp and rice paper.

The extensive utilization of natural fibers for a range of textile applications is attributed to their exceptionally brilliant mechanical and physical properties like good specific modulus, low density, toughness properties, low cost, recyclability and nontoxicity. Apart from textile and fashion industry, the natural fibers find application in various industries such as automotive, building, furniture and fiber reinforced composite products [1]. A myriad of natural fibers like jute, hemp, kenaf, ramie, flax, sisal, bamboo, coir, oil palm, etc. are used for development of polymer composites on account of their biodegradability, high performance profile, sustainable attributes, lightweight and economic viability.

Furthermore, low weight, better crash absorbance and sound insulation properties of natural fibers make them ideal choice for automotive and aerospace sectors. However, the applications of natural fibers are restricted to interior structures owing to their hydrophilic nature thereby demanding chemical treatment for improving their moisture related properties. The inherent property of moisture absorption on exposure to different temperatures and humidity conditions by natural fibers presents a key challenge for their usage in different environmental conditions [2–4].

Natural fiber composites are composite materials comprising reinforcing fibers derived from renewable and carbon dioxide neutral resources like wood or plants. NFCs find application in molded articles that demand moderate strength for acceptable performance like large diameter piping, equipment housings, roofing for economical or low-budgeted housing. A rapid drift from oil derived polymers and mineral reinforced materials to sustainable alternatives has fostered automotive and packaging industries to start utilizing natural fiber composites in their designs. Accordingly, natural fiber composites are serving as an energy efficient and sustainable alternatives replacing traditional materials such as metals, polymeric resins and reinforcement fibers. A worldwide clamor for green products and thus upsurge in sustainable alternatives has been witnessed as a result of diminishing petroleum reserves worldwide, exorbitant prices of petroleum and high disposal costs of petroleum-based composites along with inability of decomposition of some petroleum based composites. Contrastingly, natural materials outshine the petroleum based products in being renewable, inexpensive, biodegradable and eco- friendly.

The demerits associated with glass fibers have prompted the emergence and wide spread acceptance of natural fibers for making composites to be suitable in automotive, furniture, packaging and building industries. Although natural fiber composites offer innumerous advantages such as lower cost, lightweight composites, biodegradability and renewable sources, however disadvantages associated with natural fibers like variations in fiber geometry and physical properties, lower mechanical properties, poor interfacial adhesion and incompatibility with hydrophobic matrix resin systems cannot be undermined.

The challenges ahead in design and manufacturing of natural fiber reinforced composites for varied applications lie in overcoming the aforesaid demerits associated with natural fiber composites.

The next section of the chapter shall discuss the fiber properties suitable for textile applications and the broad fiber classification.

2. Fiber properties for textile applications

The properties which qualify the textile fibers to be used for textile applications are being discussed in detail in the following section:

2.1 Length and appearance

Length of fiber influence their selection for fabric production. Short staple natural fibers like cotton, jute, wool impart rough look to fabrics made from these fibers. However, fabrics made of nylon, polyester filament yarns exhibit smooth and shiny look. Moreover, short wavy fibers like cotton and wool have affinity for dirt particles while long, continuous filament fibers like silk or synthetic fibers are easy to clean and do not get as dirty as their natural, staple counterparts. Accordingly, wool, jute fibers are used for apparels and fashion accessories where designer aims to explore rough, crimpy fabric appearance as the design feature while accessories like scarves, stoles which demand sheen and smooth, soft look and feel, utilizes silk or filament fiber yarns and fabrics in their designing.

2.2 Moisture absorption

The fibers vary in their affinity for water molecules with natural fibers like cotton inherently hydrophilic while synthetic fibers like polyester, nylon exhibit hydrophobicity or low moisture absorption. The moisture absorption property of fibers determine their selection for apparels and accessories suitable for a particular season.

Summer wear apparels and accessories should provide rapid sweat absorption from wearer's skin and thus utilizes hydrophilic fibers like cotton. Fabrics made of synthetic fibers, being hydrophobic do not soak sweat and are uncomfortable in summers. A special consideration of fiber type is important while designing textile end products like gloves, socks which are in intimate contact with skin where sweat accumulation may lead to dampness and discomfort to the wearer.

2.3 Thermal conductivity

The ability of the fiber to conduct heat away from the wearer's skin is yet another crucial factor influencing the selection of fiber for a particular end use. Cotton and rayon exhibit high thermal conductivity and have the ability to conduct heat from body, lower the body temperature and providing cool feel next to skin, Accordingly, fibers with higher thermal conductivity are preferred for accessories intended for summers like driving gloves, socks, stockings. However, the requirements in winter are quite contrasting with such material selection that can provide thermal insulation and prevent the body heat to escape out. Therefore, fabrics made of wool, acrylic and synthetic fibers, being poor heat conductors of heat aid in keeping the wearer warm in winters.

2.4 Strength

The strength of fibers determine the ease of washing of fabric made from the chosen fiber. Fibers have property of strength variation in dry and wet states. Fibers like wool, silk, rayon loose strength in wet state while cotton and synthetics exhibit high strength even when wet enabling easy laundering of fabrics made from such fibers. Therefore, cotton and other synthetic fibers should be preferred for daily wear apparels and accessories which require frequent cleaning and laundering. However, occasional wear accessories like silk scarves, stoles are made of delicate fabrics prone to strength loss on repeated washings and need dry cleaning and less frequent washing cycles.

3. Classification of textile fibers and textile based applications

The various types of fibers used in the textile industry exhibit their own individual, unique properties. The properties fibers display, depend largely on their source of origin. The fiber properties desirous for textile applications have been discussed so far. This section is devoted to elaborate classification of fibers based on their nature and origin, botanical, zoological or chemical name, moisture absorption properties, thermo plastic behavior and utility. The chapter will mainly focus on origin based fiber classification.

3.1 Classification based on origin

The primary classification of fibers is based on their origin and thus can classified as natural and synthetic fibers.

3.1.1 Natural fibers

Natural fibers, as the name suggests are obtained from nature. The natural sources for these fibers can be plants, animals or minerals (**Figure 1**). **Figure 2** shows the classification of vegetable based cellulosic fibers. The fibers obtained from plants include

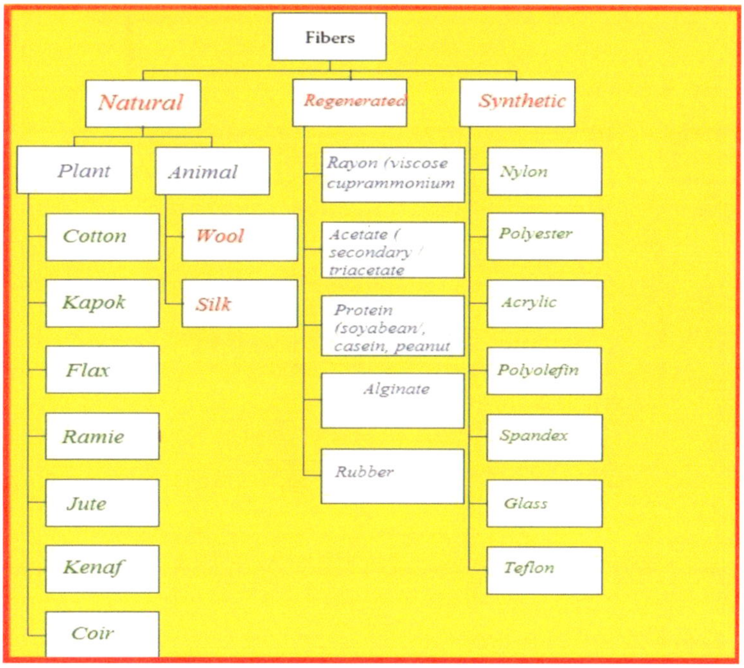

Figure 1.
Classification of fibers.

cotton, kapok (from seeds), sisal, banana, pineapple (obtained from leaves) and bast fibers like jute, flax, hemp, kenaf, ramie, etc. (obtained from plant's stems). Cellulose fiber extensively used in textile industry include cotton, linen, flax, hemp and jute. A variety of fashion ensembles ranging from cotton canvas satchels, tote bags, tapestry luggage to sports socks, sneakers, bandanas, handkerchiefs, scarfs, stoles, hats and caps uses cotton as the major raw material.

The salient features of cotton fiber that makes it suitable for a range of apparels and accessories include its softness, breathability, moisture absorbency and temperature regulation property. The natural textured surface of fiber in addition to its moisture absorption provides comfortable feel and breathability next to skin. Accordingly, the fiber is preferred for applications where rapid water absorption is the prime concern like terry-cloth apparel such as beach coats. Moreover, cotton is versatile, easy to care and handle fiber that is not just static free, hypoallergenic, and pill free but also has the ability to retain its original feel and color. The above stated attributes make cotton year round fiber suitable in both warm as well as cold weather.

Kapok belongs to the Bombacaceae family and is generally cultivated in tropical regions. Kapok seeds are enclosed within the fiber. The cellulosic fiber features yellowish or light brown color, light weight and hydrophobicity. Kapok fiber is preferably utilized as adsorption, oil absorbing, buoyancy material, apart from being used as biofuel and reinforcement material.

Flax is cellulosic fiber in crystalline form featuring a length of 90 cm and diameter of 12–16 μm. Flax is mainly cultivated in Canada, Netherlands, Belgium and France. The stem of plant L. usitatissimum is the fiber source. The fiber extraction is accomplished by two processes namely retting and scorching to alter the fiber properties. The enzymatic application during retting process causes pectin degradation thereby

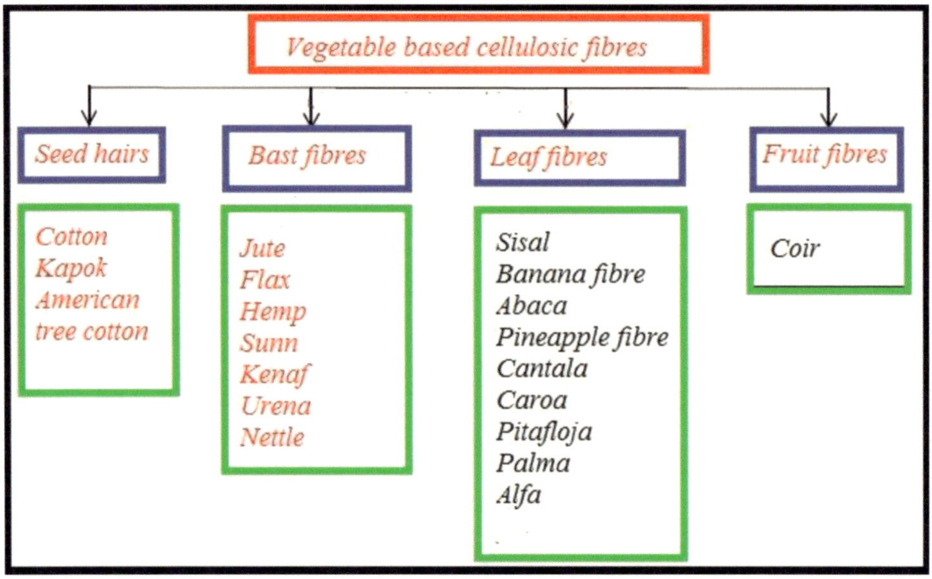

Figure 2.
Classification of vegetable based cellulosic fibers.

resulting in separation of fibers. Flax fiber is extensively used for production of linen besides being used in furniture, home textiles and interior decor items.

Ramie fiber is herbaceous perennial plants extracted from the plant steam, is extensively grown in China, Japan and Malaysia. Ramie is considered to be one of the fast growing, strongest and longest among natural bast fibers featuring a height of 1–2 m. Ramie fiber is processed in a similar manner as linen from flax. The fiber finds wide range of applications such as apparels like sweaters and cardigans, upholstery, marine packing, gas mantle, fishing nets, automotive, furniture, construction, pulp, paper, agrochemicals and composites etc.

Jute is primarily grown in Asian countries like India, Bangladesh, China, and Myanmar and it takes about 4 months to grows up to height of 15–20 cm. The extraction of fibers is accomplished about 4 months after its cultivation. The fibers are subjected to chemical or biological retting process which is essential for pectin removal between bast and wood core. The chemical retting process involves the application of chemicals like $N_2H_8C_2O_4$, Na_2SO_3, etc. while biological retting involves soaking the bundled stalks in water for about 20 days enabling the removal of pectin between the bast and the wood core thereby assisting in easy fiber separation. Subsequently, the fibers are allowed to dry.

Kenaf belongs to bast fibers and are extracted from flowers, outer fiber (bast comprising 40% of the stalks dry weight) and inner core (comprising 60% of stalks dry weight). The processing of kenaf plants upon harvesting is accomplished with a mechanical fiber separator, consuming whole stalk in pulping followed by chemical or bacterial treatment of extracted fibers for their separation from the non-fibrous content such as wax, pectin, and other substances. The fiber exhibits unique properties like biodegradability and eco-friendly attributes, stiffness, strength, toughness and high resistance to insecticides. Kenaf fiber primarily find application in production of paper, rope, cords, storage bags along with textile applications. More recently, kenaf fiber is utilized for composites apart from application in furniture, construction, packaging, automotive sector.

Coir Fiber, one of the thickest natural fiber is mainly cultivated in tropical regions including India, Sri Lanka, Indonesia, Philippines, and Malaysia and is obtained from the husk of the coconut fruit.

Coir fiber in contrast to other natural fibers has higher lignin and lower cellulose and hemicellulose content and features a high microfibrillar angle thereby rendering it several valuable properties, such as strength, resilience, resistance to weathering, and high elongation at break. The salient properties of coir fiber make it a suitable candidate for a range of applications such as upholstery, ropes, mats, mattresses, agriculture, construction, brushes etc. [5–7].

Another natural fiber that is gaining interest among textile designers for designing clothing, footwear and handbags is linen. Hemp and jute are rope like fibers and exhibit coarseness compared to linen and generally preferred for apparels and accessories that demand rough texture and durability. Hemp, jute is finding applications in accessories like fashion jewelry, handbags, belts, hair accessories, footwear, bag-pack, tote, gunny-bags and mobile covers.

Jute is a plant based multicellular fiber characterized by nodes and cross markings in longitudinal view and polygon shapes in cross-section. Jute is hygroscopic in nature with moisture regain of 12–14% and its color varies from yellow to brown to gray and is good insulator of heat and electricity. Jute fiber is used for making burlap, hessian, gunny cloth which serve as raw materials for accessories like handbags, jewelry etc.

Bamboo and straw are another plant based fibers gaining popularity among fashion designers. Bamboo is wood like tropical grass while the sources for straw are wheat stalks, grasses, sisal hemp and rice paper. Summer wear hats, hair accessories, different handbags are designed from straw, bamboo or strips of bamboo woven like straw. The pliability and sturdy hand of bamboo canes make them suitable for gripping elements and handles for handbags. However, straw is generally preferred for delicate fad accessories trending in a single season which is accounted to its inferior durability and flex abrasion resistance. Repeated bending in one position may cause fiber breakage hampering the esthetic appeal of accessory and rendering it useless.

The fibers such as wool (obtained from sheep) and silk (obtained from silk worm cocoons) are protein based fibers and finding applications in a variety of accessories. Pashmina (cashmere or cashmere/silk blend), wool and silk are used for designing exclusive and high priced stoles, scaves and other winter wear accessories. Pashmina is obtained from the underbelly of the Capras goat found in India's Himalayan mountains. Wool is easily distinguishable from hair or fur fibers showcasing crimpy appearance and elasticity. Fabrics made from wool fiber exhibit durability owing to tear and snag resistance and anti-pill properties. Moreover, wool fabrics provide easy care and handling properties, drapes stunningly, and maintains resilience in wet condition. The most crucial property of wool fiber of providing insulation by holding air layer next to the skin makes the fiber a preferred choice for winter wear apparels and accessories. Henceforth, winter wardrobe is incomplete without the inclusion of wool based hats and scarfs.

Another natural protein fiber namely silk is composed of fibroin and is the product of insect larvae that forms cocoon. The most exclusive variety of silk is obtained from the cocoons of the larvae of the mulberry silkworm *Bombyx mori*. Silk is considered to be the most luxurious of all fibers characterized by unique natural luster. Triangular prism like structure of the silk fiber is responsible for its shimmering appearance allowing silk fabric to refract incident light at different angles, thereby producing varying color effects. The silk fiber has an additional advantage of being one of the strongest natural fibers and offers remarkable abrasion resistance with many years of service to wearer.

Furthermore, the fiber provides extremely soft texture, is elastic and displays the ability to retain its original shape with minimal shrinkage.

The unique properties of this protenecious fiber available in filament form make it the most anticipated option for apparel and accessory designers to render a loyal, pleasing look to end products. The fabrics produced from this luxurious fiber however, need special care as far as dry cleaning, washing and pressing is concerned. Manual washing of silk apparels and accessories is usually recommended unless the end products have been processed for machine laundering. Moreover, the exorbitant prices of silk fiber prompt designers to look for cheaper substitutes that resemble silk, such as polyester and nylon microfibers.

Vegetable fibers like ramie, jute and hemp in contrast to hair fibers are economical and are characterized by rigid feel, coarseness and brittleness and thus find application in designing textile products where strength, abrasion resistance and rough texture are the requisites. The fibers are thus generally preferred for designing lower priced accessories like fashion jewelry, belts, handbags. Asbestos is the natural mineral fiber known for its fireproofing and insulating properties. Accordingly, the fiber was utilized for flame retardant protective clothing. However, the fiber has limited application in textile arena in recent times owing to its carcinogenic nature and hazards to lungs by continued inhalation of asbestos fiber.

3.1.2 Man-made fibers

Man-made fibers also referred to as synthetic or artificial fibers are the fibers that are developed by mankind to meet the ever increasing demands of fibers in textile and fashion industry since the sources of natural fibers are on the verge of depletion. Moreover, the stringent animal rights discourage slaughtering of animals for their skin, fur and hair fibers. Therefore, the man-made fiber industry is bound to grow tremendously. The man-made fibers are manufactured with the aim of cutting down costs associated with natural fibers and at the same time achieving desirable properties like high strength, abrasion resistance, soft feel, drape ability and varied textures. Manmade fibers are classified into regenerated, synthetic and miscellaneous inorganic fibers on the basis of raw materials employed for their manufacture. Regenerated fibers like viscose rayon, cuprammonium rayon, acetate, triacetate, casein, rubber are the man-made fibers that belong to cellulosic group and are produced using natural polymers i.e. cellulosic base materials thus requiring minimum chemical steps [1–4].

Acetate it has trade name of Ariloft, Chromspun had a luxurious feel and appearance, is available in a wide color range, drapes extremely well, is relatively fast drying and is resistant to shrinkage, mildew and moth. It is used for blouses, dresses, foundation garments, lingerie, linings, shirts, slacks and sportswear.

The non-cellulosic group include fibers like nylon, polyester, acrylic, polyolefin, spandex, glass and teflon which are protein based fibers. These fibers constitute of molecules of carbon, hydrogen, nitrogen and oxygen derived from petroleum and natural gas. The fibers manufactured by synthesizing various chemicals like the petroleum products are classified as synthetic manmade fibers. Natural raw materials are not required as a base material for their manufacture unlike the regenerated ones.

Nylon fiber belonging to this group is manufactured using hexamethylenediamine and adipic acid while polyester uses dimethyl terephthalate and ethylene glycol in its production. The process basically involves conversion of chemicals into fiber forming substances which can be drawn into filaments of required coarseness or fineness.

Nylon is known for its exceptional strength, abrasion resistance and easy care properties making it desirable fabric for hosiery, blouses, dresses, lingerie, underwear, raincoats, ski wear and suits.

Acrylic is also referred to as Acrilan, Orlon and Zefran. The fiber is characterized by its soft feel, warmness, wool like texture and light weight. Moreover, the acrylic fiber fabrics display considerable strength, resilience, drying ability, and are resistant to moths, sunlight, oil, and chemicals. Therefore, fabrics made of acrylic fiber namely fleece, fake furs, jerseys find application in winter wear and knitted accessories along with apparels like dresses, infant wear, knitted garments, skirts and sweaters. Modacrylic, another synthetic fiber is used for fabrication of fleece and knit pile fabrics. The fabrics owing to their softness, resilience, abrasion and flame resistance, drying ability, and shape retention, are preferred for deep-pile coats, linings, simulated fur accessories and hair accessories.

Olefin is another organic man-made fiber that exhibits excellent colorfastness, extreme strength, resistance to mildew and perspiration and is thus, is garnering designer's attention for designing of knitted accessories, winter wear and sportswear apparels and accessories.

Another fiber that has taken the apparel and accessory industry by storm is rayon fiber which offers multitude advantages of being soft, comfortable and easily dyeable. Consequently, the fiber is preferred choice for a variety of apparels and accessories like sportswear, blouses, jackets, lingerie, rainwear, shirts, scarves, stoles and ties.

Spandex Produced extensively by Dupont as Lycra, its major advantage is its ability to be stretched 500 percent over and over without breaking, always returning to its original length. It is lightweight, more durable than rubber, and resistant to body oils, it is used wherever stretch is required such as in athletic apparel, bathing suits, foundation garments, ski pants and sportswear.

Miscellaneous inorganic group comprises of fibers which use substances like metal and glass in their manufacture. The malleability and ductility of the inorganic fibers render them usable for textile applications. However, high cost and technical difficulties offers hindrance to wide spread acceptance of the fibers [5, 8].

4. Natural fibers as reinforcement for composites materials

There has been an upsurge toward attainment of superior mechanical and tribological properties for varied applications by replacing existing materials with more advanced, innovative and sustainable materials. Accordingly, monolithic materials are fast being replaced by materials like glass, carbon, aramid fibers for sporting, aerospace, automotive and construction sectors. Nonetheless, the non-biodegradability, non-renewability and energy intensive production processes render the materials unsustainable and thus doomed to be hazardous to environment. Consequently, natural fiber reinforced composites with potential to replace the synthetic fibers are gaining momentum to avert the deleterious impact of conventional, unsustainable materials. Accordingly, a gamut of natural fibers obtained from fruits, seeds, leaves, stem, animals, etc. are being explored for their viability in composites. The composites are tailor-made materials possessing unique qualities which can be altered by variation in reinforcement and matrix phase. The salient features of natural fibers particularly their low density (1.2–1.6 g/cm^3 compared to glass fiber with density of 2.4 g/cm^3) make them preferred choice for light weight composites in contrast to synthetic fibers. Consequently, there is an upsurge in the demand of natural fibers based composites commercially in gamut of industrial establishments such as automotive interior linings (roof, rear wall, side panel lining), furniture, construction, packaging etc. Lightweight composites primarily utilize natural fibers like hemp, jute, sisal, banana, coir, and kenaf for their production process. Animal hair fibers owing to good mechanical properties such as ductility of 20% elongation on the average and strength to failure of 250–300 MPa on the average are other suitable candidates for myriad of textile and non-textile applications. Likewise, coir fibers exhibit high impact strength/resistance.

The modification of properties of the natural fibers by different chemical treatments and blending them with polymers and other synthetic materials is accomplished to enhance the properties of the natural fibers and in turn the properties of hybrid composites. Hybridization involves combination of fillers and natural fiber resulting in enhanced mechanical properties of the composites. A number of factors such as volume or weight fraction of the reinforcement, fiber alignment, distribution, orientation and aspect ratio, fiber-matrix adhesion, usage of additives, and chemical treatment of fibers affect the mechanical performance of fiber-reinforced composites. Natural fibers are potential sustainable candidates replacing synthetic fibers in architecture, cladding, walling, and flooring. Flax, hemp, sisal, and wool are used in car interiors and exteriors like in Mercedes-Benz components. The coir/polyester-reinforced composites find application in the mirror casing, paperweights, voltage stabilizer cover, projector cover, helmet etc. Likewise, rice husk fiber, cotton, ramie, jute fiber, kenaf are increasingly being used in clothing, fishing nets, packing materials, ropes, sewing threads [5, 6].

The following section will discuss salient features of natural fibers suitable for composites and varied application areas:

The source of natural fibers include a variety of plants and animals (hair, chicken feather). The plant based fibers constitute of cellulose, lignin, hemicellulose, pectin, waxes, and water-soluble substances. **Figures 3** and **4** shows the various natural fibers suitable as composite materials and other associated sectors.

Cotton (Gossypium) is considered to one of the most crucial agricultural crop holding lion's share of market as far as textile industry is concerned. The fiber belongs to the sub-tribe Hibisceae and family of Malvaceae. The cotton cultivation is generally practiced in tropical and subtropical regions, with China being the largest cotton producer followed by India and the United States. Upland cotton (*Gossypium hirsutum*) and pima cotton (*Gossypium barbadense*) are the most popular and extensively utilized cotton species. The comfort, moisture absorbing and hydrophilic attributes make it a preferred choice for summer wear clothing and accessories. More recently, the potential of cotton for development of composites for industrial applications is being explored.

Silk (Bombyx mori) is an animal based natural fiber extracted from cocoons of silkworms. The major producers of silk include China, South Asia, and Europe. The exceptionally brilliant properties of fiber such as good mechanical strength, extensibility and compressibility makes it a preferred choice as luxury material for exorbitant, high end range of apparels and fashion items.

Hemp belongs to plants species and is generally cultivated in European and Asian regions. The fiber is approximately 2 cm in diameter and grows to a height of 1.2–4.5 m. Hemp comprises of inner layer surrounded by outer core of bast fiber

Figure 3.
Natural fibers.

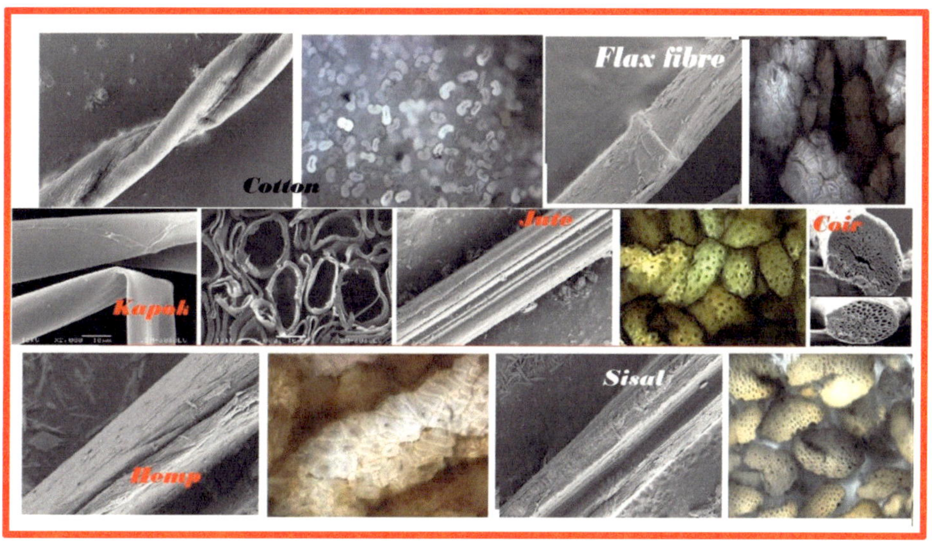

Figure 4.
Logitudinal & cross-sectional view of natural fibers.

attaching to the former by glue-like substance or pectin. The harvesting of hemp fiber is followed by separation of woody core from bast fibers through mechanical process. Thereafter, the separated woody core is cleaned for obtaining the required core content and may be cut to the desired size. The subsequent processing of separated bast fibers result in yarn or bundle formation. Hemp fiber finds application in range of textile and non-textile applications such as sustainable apparels, bags, ropes, garden mulch, fabrication of composites, building material and animal beddings.

The flax fiber extraction is accomplished by two processes namely retting and scorching to alter the fiber properties. The enzymatic application during retting process causes pectin degradation thereby resulting in separation of fibers. Flax fiber is extensively used for production of linen besides being used in furniture, home textiles and interior decor items.

Bamboo is also referred to as natural glass fiber owing to fiber alignment in the longitudinal direction. Bamboo is available in the dense forests of China and features as many as 400 species. The par excellence properties of fiber such as light-weight, low cost, high strength, and stiffness renders it suitable as reinforcement in polymeric materials, building and bridge construction, bridges, traditional boats, etc.

Pineapple Leaf fiber is one of the abundantly cultivated leaf fiber obtained from crop waste after cultivation of pineapple. The short tropical plant features a height of 1–2 m and 20–30 clustered leaves. Pineapple leaf fibers are multicellular and lingo-cellulosic. The fiber exhibits good mechanical properties thereby making is a preferred choice for varied applications such as automobiles, textile, mats, construction, conveyor belt cord, air-bag and advanced composites.

Sisal is mainly grown in Brazil and South Mexico. The fiber comprises of the rosette of leaves and grows up to a height of 1.5–2 m. Sisal fiber on account of its good mechanical properties is extensively utilized for a range of textile and non-textile applications such as fiber core of the steel wire cables deployed in elevators, automotive sector, shipping industry, civil constructions etc.

5. Natural fibers composites—Some salient features and associated challenges

Fiber reinforced polymeric composites consist of reinforcement fibers held by surrounding resin matrix. Continuous filaments or short fibers comprise the reinforcement fibers. Reinforcement preforms like nonwoven mats, yarns, fabrics and 3D fabrics are obtained from fibers.

The biological origin of at least one of the components of natural fiber composite make them potential sustainable candidates with long textile fibers (e.g., flax, hemp, kenaf, jute, ramie and sisal), short fibers (wood fibers, by-products from long fiber processing, and recycled fibers) and fiber fibrils being used as reinforcement. Biomaterials like various epoxidized plant oils and soy protein serve as matrix materials. The fabrication of natural fiber composites is accomplished in a similar manner as manufacturing methods namely resin transfer molding (RTM), vacuum infusion, compression molding, direct extrusion and compounding, and injection molding deployed for conventional composites comprising thermoset matrix and thermoplastic composites. The pre-treated process of fiber and manufacturing techniques of composites primarily govern the properties of natural fiber reinforced polymer composites. Composites with diverse properties can be obtained by varying the manufacturing techniques and the constituents of composite materials. Thus it can be inferred that precise selection of fibers, matrix, additives and manufacturing method enables tailoring the properties of natural fiber composites for varied applications.

Natural fiber composites offer several environmental benefits in contrast to their synthetic counterparts as the former result in less pollution during fabrication, lower fuel consumption and CO_2 emissions during transport to the constructions sites and lastly, considerable reduction of the disposal and energy-consuming disposal efforts.

Natural fiber composites owing to replacement of synthetic materials by bio-base and renewable sources are classic examples of sustainable resources for industrial applications. The materials exhibit the potential of being cost effective for identical structural characteristics, can be grown in controlled facilities or farms and can substantially bring down the carbon footprints. The problems posed by synthetic fibers and resins in their disposal accounts for approximately 20% of the total landfill space thereby fostering ardent replacement of synthetic composites with natural fiber composites.

Nevertheless, the applications of natural fibers are restricted to interior structures owing to their hydrophilic nature thereby demanding chemical treatment for improving their moisture related properties. The variability in fiber properties namely apparent variability and actual property variability presents a challenge in their usage as reinforcement in composite materials. Apparent variability arises due to experimental methods, measurement and testing techniques while the actual property variability is the inherent variability present in fiber. Moreover, the structural defects in the fibers leads to fiber deformation which also restrict the usage of natural fibers.

The inherent property of moisture absorption on exposure to different temperatures and humidity conditions by natural fibers presents a key challenge for their usage in different environmental conditions [2, 3]. The presence of hydrophilic group affects the interfacial bonding between polymer matrix and the fiber in composite structures owing to hydrophobic characteristics of matrix. The interaction between fibers and polymer matrix can be optimized by chemical treatment of natural fibers by reduction in hydroxyl functional groups on the fiber surface thereby increasing surface roughness and the interfacial interaction between the matrix and the fibers [6, 7].

6. Conclusions

The significance of natural fibers to fulfill the basic human needs of clothing and shelter has been well established since time immemorial. However, with the advent and upsurge in the popularity and demand of synthetic fibers, the global consumption of natural fibers became archaic. Nevertheless, the world wide clamor for sustainable and eco-friendly approaches in textile supply chain and depletion of petroleum resources have stimulated the usage of natural fibers thereby replacing synthetic fibers with sustainable natural fibers.

Cotton holds lion's share of market as far as textile applications are concerned but the fiber cannot be considered sustainable owing to massive utilization of water, pesticides, fertilizers and toxic chemicals thereby leading to environmental and economic distress. Apart from cotton, some resource efficient fibers namely hemp, flax, organic cotton, bamboo, jute, kenaf, ramie, sisal, coir are replacing cotton in varied textile applications and for development of polymer composites on account of their biodegradability, high performance profile, sustainable attributes, lightweight and economic viability.

Natural fiber composites are composite materials comprising reinforcing fibers derived from renewable and carbon dioxide neutral resources like wood or plants. NFCs find application in molded articles that demand moderate strength for acceptable performance like large diameter piping, equipment housings, roofing for economical or low-budgeted housing. Natural fiber composites owing to replacement of synthetic materials by bio-base and renewable sources are classic examples of sustainable resources for industrial applications. The materials exhibit the potential of being cost effective for identical structural characteristics, can be grown in controlled facilities or farms and can substantially bring down the carbon footprints. Natural fiber composites offer several environmental benefits in contrast to their synthetic counterparts.

However, the applications of natural fibers are restricted to interior structures owing to their hydrophilic nature thereby demanding chemical treatment for improving their moisture related properties. Undoubtedly, natural fibers are potential sustainable candidates to replace synthetic fibers for myriad of textile and non-textile applications.

Author details

Yamini Jhanji Dhir
Department of Fashion and Apparel Engineering, TIT&S, Bhiwani, Haryana, India

*Address all correspondence to: yjhanji@gmail.com

References

[1] Kılınc AÇ, Seydibeyoglu MO Woodhead. Natural fibres Cambridge. In: Fiber Technology for Fiber-Reinforced Composites. Cambridge: Woodhead Publishers; 2017

[2] John MJ, Molaba TP. Mechanical properties and water sorption of chemically modified natural fiber-based composites, in encyclopedia of materials: Composites. Encyclopedia of Materials: Composites. 2021;**3**:159-167

[3] Anish M. Varghese, Vikas Mittal, Polymer composites with functionalized natural fibers in Biodegradable and biocompatible polymer composites. Biodegradable and biocompatible polymer composies. 2018;**3**:157-186

[4] Kozłowski RM, Mackiewicz-Talarczyk M. Introduction to natural textile fibres, handbook of natural fibres types, properties and factors affecting breeding and cultivation, volume 1 in woodhead publishing series in Textiles. 2nd ed. Elsevier. 2020;**1**:1-13

[5] Everitt NM, Aboulkhair NT, Clifford MJ. Looking for Links between Natural Fibres' Structures and their physical properties. Conference Papers in Science. 2013;**2013**:1-13. DOI: 10.1155/2013/141204

[6] Sfiligoj Smole M, Hribernik S, Stana Kleinschek K, Kreze T. Plant fibres for textile and technical applications. Advances in Agrophysical Research. DOI: 10.5772/52372

[7] Girijappa T. Natural fibers for eco-friendly composites of these fibers. Frontiers in Materials. 2019;**27**:1-15. Available from: www.frontiersin.org

[8] Adnan M, Muhammad A, Sarwar I. Sustainable and environmental friendly fibers in textile fashion—A study of organic cotton and bamboo fibers master thesis, University of Boras Report No. 2010;**3**:53-60

Section 2

Properties of Natural Fibers

Chapter 2

Basic Functional Application for Natural Fibers and Types

Ramratan Guru, Anupam Kumar and Rohit Kumar

Abstract

Natural fiber has a more long history and these are very important in a wide range of applications in the textile sector. Basically, natural fibers have biopolymers and natural fibers are basically made from either plant or animal-sourced. The plant-based natural fibers are major constituent of cellulose content and animal-based natural fibers are comprised of proteins. Nowadays, more used around the world for the plant-based natural fibers to bioplastics, biocomposites materials in automotive industries. These make product are low cost, low density, low manufacturing energy consumption, and more biodegradable. The ever-growing environmental, ecological, and economical concerns lead to increased acceptance of natural fibers in every area of conventional synthetic material application. This is due to biodegradability, nontoxicity, combustibility, easy availability, nonabrasiveness, and good specific strength. The present study focuses on the functional application aspect of natural fibers, basically an identification of fiber, classification and application of fibers process parameters.

Keywords: natural fibers, classification and application of fibers process parameters

1. Introduction

Basically fiber is defined as a unit of substance characterized by flexibility, fineness, length, and thickness. In the order of textile, the fiber basically used have should be sufficiently high-temperature stability, strength, elasticity, and moisture performance. Generally, textile fibers are basically of two categories: natural sources and man-made fibers. They are fibers from natural sources like plants and animals etc. and do not require fiber formation, are categorized as natural fibers. The natural fibers are basically of two categories like cellulose fiber such as flex, hemp, cotton, mineral fiber, and another classes protein fibers are such as silk and wool [1, 2].

Man-made filaments are filaments in which either the introductory chemical units have been formed by chemical conflation followed by fiber conformation or the polymers from natural sources have been dissolved and regenerated after passage through a spinneret to form filaments. Those filaments made by chemical conflation are frequently called synthetic filaments, while filaments regenerated from natural polymer sources are called regenerated filaments or natural polymer filaments [3]. In other words, all synthetic filaments and regenerated filaments are man-made filaments, since man is involved in the factual fiber conformation process [4, 5]. In discrepancy, filaments from natural sources are handed by nature in ready-made form. Basically, man-made fibers contain polyesters, acrylics, polyamides (nylon), vinyls, elastomeric fibers, polyolefins, while the regenerated fibers include rayon, cellulose acetates, the regenerated proteins, glass, and rubber fibers. Basically, this article has the main purpose of all types of textile fibers, gives brief knowledge with specification facts.

2. Basic properties of a textile fiber

In this session, we have described the properties of all kind of fibers, which are commonly viewed as important aspects.

2.1 Fineness

The fineness is very important part of each fiber. The thickness of fiber can be known from its width, diameter, and sectional area. But there are very few fibers that have a completely round sectional area. So it is difficult to get the perfect answer. Therefore, there is a number that shows the ratio of weight against a fixed length or vice versa, the ratio of length against a fixed weight. For example, fineness is indicated by Denier, Tex, or yarn count. The excellence of fiber quality is evaluated from its fineness.

2.2 Length/diameter

Length is an important characteristic that defines the usefulness of a textile fiber. The length should also be many times its diameter. In general, this would mean that when one talks of fiber length in terms of a few centimeters it has to be a few microns of fiber diameter. The staple length of spinnable fibers is generally not less than 18 mm. Fibers below 5 mm are just not integrated into the yarn. In the case of filament fiber this ratio would be very large and perhaps irrelevant. The cut staple length depends on the spinning system to be used and the fiber it is blended with in case of blends [6].

2.3 Strength

It is essential that the fabric should be durable enough. For durability, the fabric must be strong enough. The strength of the fabric is more influenced by the strength of the fiber present in the cloth. Basically it indicated strength to resistance constant by fibers, yarns form, and cloths to breakdown when energy is applied to them. The basically is strength parameters like bending, tensile and bursting, etc. according to the direction and application power.

2.4 Elasticity

Basically, elasticity performance depends on capability of the garments to the material to area imaginative nature after being deformed by the use of strength. Elasticity or elastic recovery is generally influenced by the extent of stretch, during time which material is kept in its stretched condition, and the time to recover.

2.5 Uniformity

It is essential that there should be limited variations in length and diameter between the fibers to fiber. In other words the fiber should be more uniform which will ensure uniformity in the yarn as well as in the fabric.

2.6 Spinnability

It indicated that the individual fibers must be capable of being spun into a yarn and then fabric with sufficient strength. For better spinnability the fiber must have better cohesiveness i.e., they must hold together to prevent slippage. The spinnability is normally used for the man-made fiber-developed procedure.

2.7 Hygroscopic property

If a fiber is left in the atmosphere, it has the properties of absorbing moisture automatically. The limit of this absorption differs according to the kind of fiber. It differs even in the same fibers according to temperature and relative humidity. Generally, the volume of moisture absorption increases along with the increase of humidity. However, the increment ratio is not always in direct proportion with the increase in humidity.

2.8 Thermal conductivity

Fibers are mostly used in raw material for clothing and the purpose of clothing is to decorate. However, the main purpose is to prevent from cold or heat (specially to prevent from clod). The amount of thermal conductivity of fibers is one of the important properties.

2.9 Resistance to chemical agents

The reaction of fibers to chemicals varies a lot according to their types. But generally, nature fibers of vegetable origin are weak in acids and strong in alkalis. Other natural fibers of animal origin are strong in acids and weak in alkalis. The man-made fiber, fibers of the cellulose series are weak in both acid and alkali, whereas synthetic fiber is stable to a certain extent in acid also and alkali also.

3. Classification of textile fibers

All the textile fiber classification is mention in below **Table 1**.

3.1 Natural fiber

Natural fibers are those, which are obtained from plants, animals, or minerals.

3.1.1 Vegetable fiber

Among vegetable fibers, the kind of fiber differs according to the part of the tree/shrub, from which it is taken.

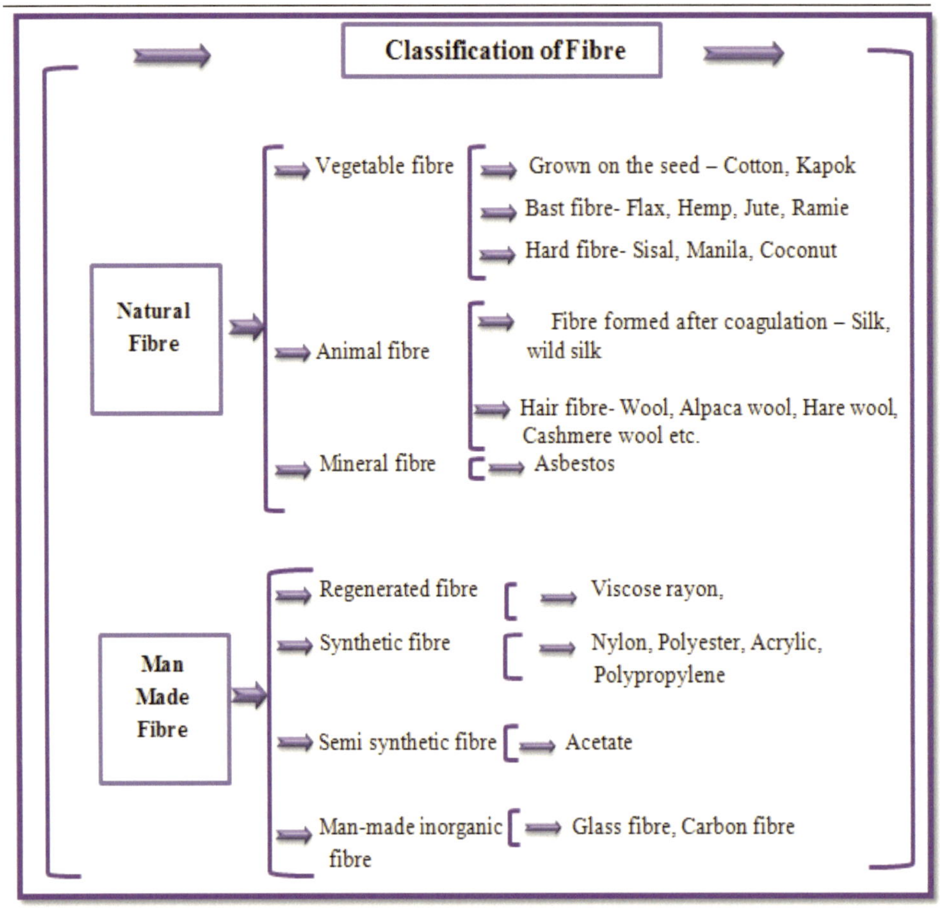

Table 1.
Classification natural fibre.

For example, collections of fiber growing on the seeds like raw cotton, Kapok, etc. collection of grown as the skin of the plant stem (bast) like flax, ramie, hemp, jute, etc. the collection of fiber from fruit shells like coir fiber (coconut fiber). The out of this most important one is raw cotton and next to it are flax, jute, Manila hemp, and ramie [7, 8].

3.1.2 Cotton fiber

Raw cotton is being used as a material for clothing for a long and its origin can be traced to 2200 BC. The cotton hair during its growth is almost cylindrical and contains a central canal called is the lumen. When remove from the seed, however the cell collapse into a flat ribbon which forms an irregular spiral band under the influence of light and air.

The cotton fibers are having different lengths. These depend upon the types of soil, weather condition, duration of harvesting the crops, etc. the length of cotton fibers are expressed in the terms of staple length. Very good fibers measure a length of 2 inches. The length of fiber also depends on the fineness of the fibers, longer the

DOI: http://dx.doi.org/10.5772/intechopen.102998

fiber finer will be the diameter. The length of cotton fiber varies from 1200 to 1300 times its width.

Strength and Elasticity: Cotton is a strong fiber among the natural fibers; the cotton gets additional strength in wet conditions. In addition to strength, cotton possesses good elasticity. Tenacity Dry 2.1–6.3 gpd and Wet 110–130% Dry.

Color: Usually cotton fibers are creamy white in color. Most of Egyptian cotton has a yellowish-brown color.

Luster: Among the important natural fibers cotton has the lowest luster which however can be improved by mercerization. To reduce the luster, delustering agents are used.

Effect of temperate: Cotton fibers ignite easily and burn to produce a smell that is obtained by burning paper or wood.

Exposure to heat: There is no visible change until a temperature of 120°C is reached. At this temperature, cotton becomes yellowish after 5 hours of exposure but shows little change in strength. At 140°C the color changes to brown and develops a serious loss of strength. Again, when heated for a few hours at 240°C, it loses its complete strength.

Moisture relation: The percentage of moisture in the fiber influences the spinning quality of fiber. Cotton cannot be spun satisfactorily in a dry atmosphere. Cotton contains 7–8.5% of moisture regain. Fabrics made from these fibers get soiled and crushed easily, but they can be washed and ironed without causing damage.

Effect of light: Cotton fibers are affected by ordinary light and more particularly by ultraviolet rays and they lose their strength to some extent. Cotton has the least resistance to exposure to sunlight.

Action of Acid: Cotton fibers are degraded by strong acids like HCl, H_2SO_4, and HNO_3. Dilute solutions of these acids if washed immediately to no harm, organic acids like acetic acid do not harm cotton fibers.

Action of Alkalis: Dilute alkalis at room temperature do not injure cotton. If cotton is boiled in a dilute solution of an alkali-like caustic soda in the presence of air, it turns yellow and loses its strength.

Dyeability: Cotton can easily be dyed using direct, sulfur, naphthol, vats, reactive and similar types of dyes.

Application for cotton fibers: It is used in summer wear, towelings, furnishings, medical textiles, canvas, sewing threads, dress material jeans and in sewing thread, etc.

3.1.3 Kapok fiber

Kapok is a silky fiber obtained from the pod of kapok tree. The botanical name is Ceiba pentranda of the family Boombacaceae. This tree is grown principally in Java, Africa, Netherlands, and South East Asia, where the soil and hot climate conditions are especially suited for its growth. The fibers are contained in the outer shell, loosely surrounding the seeds and entirely free from the cell. The Kapok fiber has a hollow structure with an external radius of around 8.25 (±4) mm, internal diameter around 7.25 (±4) mm, and length around 25 (±5) mm. Combined with the specific material density of 1.3 g/cm^3.

Application: Kapok fibers are moisture-resistant, resilient, soft, and brittle. This fiber is not suitable for spinning but it is very much suitable for stuffing. So, this fiber is mostly used in jackets, sleeping bags, insulating materials, and upholstery.

3.2 Bast fiber

Basically are the bast fiber category in Jute, Flex, Hemp, Ramie, etc. [9].

3.2.1 Jute fiber

Jute fiber is moderately strong, lustrous & yellowish-brown in color. It tends to disintegrate in water and has poor elasticity. However, this rigidity becomes virtuous. It is our best bagging material. Jute is difficult to bleach and it cannot be made pure white. It is the most important among all bast fibers. It is 2rd only to cotton in terms of crop polymer. It is easy to be spun but deteriorates when exposed to moisture. It can be converted wool-like fiber by treatment with strong caustic soda. It is highly hygroscopic with moisture regain of 13.75% and moisture content of 12% and the staple length of the fiber varies between 60 and 120 inches. Its color varies from yellow to brown. Generally, fiber is coarser and it is harsh. It is attacked by bacteria when damp. Jute is mixed with wool. Jute cloth is used for covering the cattle during winter because of its thermal insulation properties. It can be used for backing cloth for carpets and in the making of gunny bags, ropes, etc. it can be substituted for plywood also [10, 11].

Application: Jute is used for hessian sacking, backing for rugs, webbing, carpets, wall coverings, thread, canvas, matting, hammocks & beltings.

3.2.2 Flex fiber

Flax is also called as linen and many times it is called as linen, when it is turned into yarn or fabric. This is known very well around the world as the oldest of all the cultivated fiber raw materials. Flax is the bast fiber found in the stem of the plant "Linium usitatissium". The plant is cultivated in cold and humid conditions. So, the plantation is centered in cold countries. The major source of supply of flax is from the old U.S.S.R. the other countries which have flax growing areas are North Ireland, Egypt, Japan, Brazil, France, U.S.A., Australia, Canada, etc. Like jute, flax is an annual plant. The plant from which the fiber extracts grows in moist and cold conditions. The plant grows up to 160 to 170 cm in height and 1.5 cm in diameter. The tree is matured by changing its color from green to yellow. The flax fiber color is yellowish to gray, length 18 to 30 inches, elongation at a break of 2.7 to 3.5%, and moisture regain 10 to 12% [12].

Application: The flax fibers are used in household clothing fabrics, lace, sheetings, canvas, threads, papermaking, and industrial application like a fire house.

3.2.3 Hemp fiber

Basically hemp fiber is bast fiber category and similar harvesting process like for the flex fiber. These fibers are thick as compared to flax and darker color, tough to bleach process. This fiber is strong and more durable. The strands of the hemp fiber, approximate length of 6 to 8 feet and fiber length of 1.2 to 2.5 cm. The hemp fiber cross-section is polygonal shape and fiber is very stiff and surrounds considerable lignin. They are hemp fiber to produce for the coarse count cloths like sack material, rope, canvas, etc. Generally, hemp fiber is color yellowish to deep brown and moisture regains 12%. The hemp fiber is very poor elasticity recovery performance [13].

Effect of acid: Hemp is attacked by hot dilute acid or cold concentration acids which it breaks down.

Effect of alkalis: Hemp fiber has excellent resistance to alkalis.

Effect of organic solvent: It does not affect by the organic solvents.

Application: The coarse hemp fibers to produced yarn are used in woven into cordage, rope, sacking and heavy duties tarpaulins and fine hemp fibers are used for hats, shawls, rugs, towels.

3.2.4 Ramie fiber

The ramie fiber is also bast fiber categories and generally to known this fiber as a china grass. All produced fiber processes are similar to hemp fiber. The ramie fiber is a white color with more luster and good strength. This fiber is basically used for industrial application and furnishing where rough, irregular clothes are desired. The plant grows to a height of 1 to 3 meters with a diameter of approximately 8 to 20 mm thick. The plant requires a tropical climate, where the winter temperature should be above freezing. This plant is also grown in India Australia, America, Japan, Brazil, etc. Ramie is a perennial fiber, with a yield from two to five crops of fiber per year, which depends upon the soil and climate. Ramie is ready for harvesting when the lower part of the stalk turns light yellowish-brown and the lower leaves matured by turning yellow and detachable. Harvesting is done by cutting the stalks. The physical properties of Ramie fiber exhibited high tenacity, high luster, and brightness. It has resistance to heat, light, acid, and alkali, etc. The ramie fiber is moisture regain used 12% [14, 15].

Application: In general, the fiber is strong, stiff with silky luster. The ramie fiber is used for sailcloth, ropes, fishnets, paper making, and upholstery fabric.

3.3 Animals fibers

They basically are used for animal fiber like wool and silk.

3.3.1 Wool fiber

Animal hairs are obviously natural clothing material; they protect the body from wind and rain and also soften the extremes of temperature in various climates. A typical hair contains three parts the cuticular layers or epidermis, the fiber layer or cortex, and the pith or medulla. The wool is the haircut and collected from the sheep. Therefore before elaborating about the sheep wool, it is necessary to elaborate a little about the sheep. In the wool market, they are broadly classified into merino variety and crossbred variety. Sometimes only are used for comeback variety [16].

3.3.1.1 Specifications for wool fiber

Strength: As wool is the weakest of the natural textile fibers some wool fabrics may be made more durable by the use of selected grades of recycled wool, although durability is gained at the expense of texture and resiliency, and tenacity dry 1 to 1.7 gpd and Wet 76 to 97% of dry.

Elasticity: One might look upon wool's elasticity as compensation for its relative weakness depending upon the quality of wool. The fiber may be stretched from 25 to 30% of its natural length before breaking. This characteristic reduces the danger of tearing under tension and contributes to free body movements to preserve this natural elasticity. Woolen garments should be hung properly after wearing and allowed to relax sufficiently to regain their shape elongation by 25 to 30%.

Resiliency: Because wool fiber has a high degree of resiliency, wool fabric wrinkles less than some others, wrinkles disappear when the garment of fabric is steamed. Good wool is very soft and resilient, poor wool is harsh.

Drapability: Wool's excellent draping quality is aided by its pliability, elasticity, and resiliency. Wool has superior drapability than many of synthetic fibers.

Absorbency: Primarily wool tends to be water repellent. One can notice that precipitations of water on the surface of wool fibers are freely soft off. However, once the moisture seeps through the scales of the fibers, moisture is absorbed. The fibers have a degree of capillarity that will readily absorb about 20% of its weight of moisture without feeling damp. Moisture regains 16%, moisture content 13.8%.

Cleanliness & washability: Dirt tends to adhere to wool fabric. Unless thoroughly cleaned, wool retains odors, consequently, wool laundering is required.

Reaction to bleaches: The household bleached that contain sodium hypochlorite or other chlorine compounds are harmful to wool. However, bleach-containing hydrogen peroxide or sodium perborate may be used.

Application: Wools resilience bulk & ease in handling makes it the most appropriate for knitting goods. Both Woolen & worsted can be used in men's and women's suiting's, overcoats, sweaters, upholstery fabrics, and blankets.

3.3.2 Silk fiber

Silk originated from the silkworm which is cultivated in a warm shiny climate and usually employs cheap labor. The silk fabrics comprise the fabrics woven with raw silk and degummed after weaving and the fabrics woven by using the degummed silk yarn. They generally are used raw silk material to make woven cloth in white color. They are dyed on fabric surface plain color or according to consumer demand printed used. There is also a method of degumming and dyeing in which the dye is put into the degumming tank and dyeing is carried out instantaneously with scouring [17, 18].

Strength: Silk is a natural fiber. The continuous length of the filaments in thrown yarns provides a factor of strength above what is possible with short natural fibers. The smoothness of the silk filaments yarns has the inherent strength of silk along with its fine diameter for durable fabrics. The strength of silk fabric is also affected by its construction & its finish.

Tenacity dry 2.8 to 5.2 gpd, tenacity wet 75 to 95% of dry.

Resiliency: Silk fabrics retain their shape and resist wrinkling rather well. This is particularly true of the fabrics made from pure silk and from wild silk.

Absorbency: The good absorptive property of silk also contributes to its comfort in a normal atmosphere. Silk fiber can generally absorb about 11% of its weight of moisture, but the range varies from 10% to as such as 30%.

Cleanliness and washability: Silk is a hygienic material because of its smooth surface does not attract dirt, when dirt does gather; it is given up readily by washing or dry cleaning. Care should be exercised in laundering silk that always use a mild soap and strong agitation in the washing machine must be avoided as silk weakens slightly when wet.

Reaction to bleach: Strong bleaches containing sodium hypochlorite will deteriorate silk. A mild beach of hydrogen peroxide or sodium carbonate may be used with normal caution.

Shrinkage: Because of the straightness of the filament, smooth-surfaced silk fabrics have only a normal shrinkage which is easily restored by ironing.

Effect of light: Continuous exposure to light weakens silk faster than either cotton or wool. Raw silk is more resistant to light than degummed silk and weighted silk has the least light resistance.

Affinity for dyes: Generally is silk very good affinity for dyes application. It readily absorbs acid and direct dyes.

Application: Silk is famous for its lightness, luster, and beauty. It has a wide variety of uses in apparel and upholstery fabrics in terms of satin, crepe, printed crepe, voile, chiffon, damask, muslin, velvet, brocade (**Tables 2** and **3**).

3.4 The basic identification of fiber

Generally, fiber identification is of three types microscopy view, burning test, and solubility test [19–21].

S. No.	Type of fiber	Fiber moisture regain (%)	Fiber density (g/cm^3)
1.	Cotton	7.5 to 8	1.52
2.	Kapok	Above 7	1.30
3.	Jute	12 to 13.5	1.52
4.	Flax	10 to 12	1.52
5.	Hemp	12	0.83
6.	Ramie	10 to 12	1.50 to 1.55
7.	Sisal	11	1.46
8.	Coconut	8 to 12	1.18
9.	Banana	13	1.19
10.	Bamboo	12.7	1.1
11.	Wool	12 to 14	1.31
12.	Silk	11	1.34

Table 2.
Natural fiber moisture regains and density specification.

S. No.	Type of fiber	Fiber moisture regain (%)	Fiber density (g/cm^3)
1.	Nylon	4	1.14
2.	Polyester	0.4	1.38
3.	Polypropylene	0.01	0.769
4.	Polyurethane	1.3	1.0
5.	Acetate	6	1.32
6.	Viscose	11 to 13	1.46 to 1.54
7.	Acrylic	1.5 to 2	1.17

Table 3.
Man-made fiber moisture regains and density specification.

3.4.1 Microscopy view

The textile fibers, particularly the natural ones, have typical longitudinal and cross-sectional shapes and therefore can be identified by viewing them under the optical microscope. This technique cannot be used very successfully in the case of man-made fibers, except for a few because their cross-sections can be modified during production. Typical cross-sectional and longitudinal shapes of some of the fibers are given in the following **Tables 4** and **5**.

S. No.	Fiber	Longitudinal appearance	Cross-sectional shape
1.	Cotton	Ribbon like Convolutions (twist) that often change direction	Collapsed, bean-shaped, Irregular size, lumen visible
2.	Flex	Presence of cross marking and nodes. Pointed tips and smooth outline is present	Fiber bundle, fiber exhibit polygonal structure with sharp angles and small central lumen.
3.	Jute	Poorly defined nodes fiber present in bundles observation spiral elements with cross markings.	Fiber bundles with irregular outline. Fiber exhibit polygonal structure with sharp angles, outline regular in shape with thick circular.
4.	Hemp	Fiber bundle, cross markings, and nodes present. Smooth and pointed tips	Polygonal with sharply defined angles with small central lumen.
5.	Ramie	Broad ribbon like fibers and longitudinal striations, rounded tips.	Flattened structure, radical fissures, elongated lumen, and thick walls.
6.	Gummed Silk	Irregular elliptical Ribbons	Triangular with rounded corners in pairs.
7.	Degummed silk	Single, smooth, nearly structure less	Triangular cross-section with rounded corners.
8.	Tussar Silk	Flat irregular ribbons	Very elongated triangles normally separate, with rounded corners.
9.	Wool	Rough surface scales, medulla or central fiber or core	Round or nearly round, medulla may appear shaped

Table 4.
Microscopic appearances of the natural fibers.

S. No.	Fiber	Longitudinal appearance	Cross-sectional shape
1.	Polyester, Nylon and Polypropylene	Rod like smooth profile	Regular, circular
2.	Acetate, Triacetate	Distinct lengthwise striations	Irregular, serrated
3.	Acrylic	Broad, indistinct lengthwise striations	Irregular, dog bone shape
4.	Viscose	Distinct lengthwise striations	Irregular, serrated

Table 5.
Microscopic appearances of the man-made fibers.

3.4.2 Burning test

This test is basically identification for fiber smell, bead, and burning behavior performance (**Table 6**).

S.No.	Fiber	Inflame	Behavior outside the flame	Smell	Residue
1.	Cotton, Jute, Flex and Viscose	Burns quickly	Continues to burn	Burning like paper	Light gray ash
2.	Wool	Burn slowly	Self-extinguishing	Burning like hair	Crushable black bead
3.	Silk	Burn slowly	Self-extinguishing	Burning like hair	Crushable black bead
4.	Polyester	Melts. Burns slowly	Burns, drips may extinguish because of dripping	Chemical smell. Slightly sweet, chemical odor.	Hard tough light colored bead
5.	Nylon	Melts. Burns slowly	Burns, drips may extinguish because of dripping	Burning beans like	Hard tough light colored bead
6.	Acetate	Burn quickly	Continue to burn noncrushable	Acid (hot vinegar)	Hard black bead
7.	Acrylic	Burn rapidly	Continue to burn	Acid like	Irregular, hard black bead
8.	Polypropylene	Burn rapidly	Burns continuously	Burning like plastic	Hard tough tan bead

Table 6.
Burning behavior common fibers.

S. No.	Fiber	Solubility
1.	Cotton	75% H_2So_4 at room temperature
2.	Wool	Soluble in 5% NaoH at room temperature, soluble in 0.25% sodium hypochlorite solution
3.	Silk	Soluble in 5% NaoH (Hot)
4.	Nylon	Soluble in formic acid 85% and M-Cresol
5.	Polyester	Dissolves in ortho chlorophenol at room temperature, 95°C meta cresol soluble. Concentration 75% H_2So_4 at room temperature soluble polyester,
6.	Viscose	Dissolves in sodium zincate solution, 59% Sulfuric acid dissolves
7.	Acrylic	Dissolves in DMF, DMSO, Ammonium thiocyanate (70% solution at boil)
8.	Polypropylene	Dissolves in boiling Xylol, floats on water, Meta Xylene (at Boil)
9.	Acetate	Cold acetone, glacial acetic acid at 25 °C
10.	Triacetate	Solution in chloroform and methylene dichloride

Table 7.
Identification of fibers through solubility tests.

3.4.3 Solubility test

The solubility of a fiber in the specific chemical component is frequently means of fiber identification (**Table** 7).

4. Conclusion

In present study has given an overview concept of basically different types of textile fibers, classification fiber, and fiber identification performance knowledge. According to all textile fibers important characteristics view in the following conclusion are drawn:

- The overall appearance and luster of a textile can be related to the shape and light-absorbing and scattering characteristics of the individual fiber within the structure.

- They are generally man-made and natural source, both fibers are important aspects got in comfort properties like fiber fineness, strength, length, and moisture regain, etc.

- A number of fiber end-use properties in textile constructions relate to the esthetic, tactile, and comfort characteristics of the fiber.

- Textile fibers is a vast and challenging field in which required functionality can be designed by a suitable choice of raw material, fabric structure, cloths design, and finishes.

- Due to the suitable properties of fibers such as cotton, hemp, polyester, elastane, and blends of fibers and filaments, their use in comfort clothing is of paramount importance.

- Protection and safety are the important design aspects of garments a fabric, which provides comfort to the wearer by protecting it from adverse weather conditions and also enhances the performance of the selection of fiber.

Author details

Ramratan Guru[1*], Anupam Kumar[2] and Rohit Kumar[2]

1 Department of Handloom and Textile Technology, Indian Institute of Handloom Technology, Varanasi, UP, India

2 Department of Textile Engineering, Giani Zail Singh Campus College of Engineering and Technology, Maharaja Ranjit Singh Punjab Technical University, Bathinda, Punjab, India

*Address all correspondence to: ramratan333@gmail.com

References

[1] Ziabicki A. Fundamentals of Fibre Formation. New York: John Wiley and Sons; 1976

[2] Cook Merrow JG. Handbook of Textile Fibres. Welford, England: Woodhead Publisher Ltd; 1984

[3] Shenai VA. Textiles Fibres. Bombay: Sevak Publications; 1991

[4] Hamby DS. The American Cotton Hand Book. New York: Wiley and Inter Science; 1965

[5] Gupta VB, Kothari VK. Manufactured Fibre Technology. Dordrecht: Springer Science; 1997

[6] Happy F. Applied Fibre Science. New York: Academic Press; 1978

[7] Young RA, Rowell RM. Cellulose Structure Modification and Hydrolysis. New York: Wiley International; 1986

[8] Bledzki AK, Gassan J. Natural fiber Reinforced Plastics. Handbook of Engineering Polymeric Materials. New York, USA: Marcel Dekker; 1997

[9] Franck RR. Bast and Other Plant Fibres. Boca Raton, FL: CRC Press; 2005

[10] Mohanty AK, Misra M. Studies on jute composites: A literature review. Polymer-Plastics Technology and Engineering. 1995;**34**(5):729-792

[11] Atkinson RR. Jute Fibre to Yarn. New York: Chemical Publishing Co., Inc.; 1965

[12] Yan L, Chouw N, Jayaraman K. Flax fibre and its composites: A review. Composites Part B: Engineering. 2014;**56**:296-317

[13] Shahzad A. Hemp fiber and its composites: A review. Journal of Composite Materials. 2012;**46**(8):973-986

[14] Ugbolue SCO. Structure/property relationships in textile fibres. Textile Progress. 1990;**20**(4):1-43

[15] Kucharyov MS. Properties and modification methods for vegetable fibers for natural fiber composites. Tekst. Prom. Journal of Applied Polymer Sciences. 1993;**23**:8-9

[16] Von Bergen W. Wool Hand Book. New York: Inter Science; 1963

[17] Gulrajani ML. Silk Dyeing, Printing & Finishing. New Delhi: IIT Delhi; 1988

[18] Peter RH. Textile Chemistry Vol. I, Chemistry of Fibres. Amsterdem: Elsevier Publishing Co; 1963

[19] Catling D, Grayson J. Identification of Vegetable Fibres. London: Chapman and Hall; 1982

[20] Mishra SP. A Textile Book of Fibre Science and Technology. New Delhi: New Age International Publishers; 2010

[21] Gohl EPG, Vilensky LD. Textile Science. New Delhi: CBS Publishers & Distributors PVT. LTD; 1999

Chapter 3

Bast Fiber Textiles Addressed Improvement of Human Life

Malgorzata Zimniewska and Barbara Romanowska

Abstract

Cannabis sativa L. and *Linum usitatissimum L.* belong to fibrous plant family delivering textile fibers located in their bast of stalk. This chapter covers discussion about flax and hemp fibers properties and processing based on authors' finding and available literature. The authors will present research on flax and hemp fibers bioactivity in relationship with their chemical composition, which is strongly related to the selected method of fiber processing, including methods of fiber extraction in light of their effect on fibers antioxidant and antibacterial activity. Human-ecological features of linen/hemp textiles, including clothing effect on human physiology, are described. The case study of functional clothing preparation based on the bioactivity of bast fibers will be presented. This chapter delivers knowledge about complex factors of human-ecological performance of flax and hemp, which have a significant effect on the improvement of human life, including comfort, well-being, and health-supporting performance. The environmental approach of bast fibers in terms of contribution to green planet protection is shortly discussed. Collected literature and authors' findings allowed to prove the positive effect of bast fibers textiles on the improvement of human life in terms of everyday wearing of clothing as well from the viewpoint of environmental impact, which is in line with the European Green Deal strategy.

Keywords: flax, hemp, fiber processing, fiber properties, bioactivity, antioxidant activity

1. Introduction

Cannabis sativa L. and *Linum usitatissimum L.* belong to fibrous plant family containing fibers occurring as bast of the stalk. Both the fibrous plants have gained importance in the face of bioeconomy development in the whole world resulting from pro-ecological character of the flax/hemp value chains as well as a growing preference for natural raw materials, which are desired for bioproduct manufacture in many sectors.

Observed climate change and environmental degradation impose taking measures that will protect the globe for future generations [1]. The European Green Deal (EGD) set up three main objectives to counteract negative phenomenon:

- no net emissions of greenhouse gases by 2050,
- economic growth decoupled from resource use,
- no person and no place left behind.

One of the assumptions of the European Green Deal is the improvement of the well-being and health of citizens by providing:

- fresh air, clean water, healthy soil, and biodiversity,
- longer lasting products that can be repaired, recycled, and reused.

C. sativa L. and *L. usitatissimum L.* plants, delivered by them raw materials as well as flax/hemp value chains and bast-fiber-based bioproducts are fully in line with the EGD measures. Their impact on environment and agriculture is discussed in many scientific articles [2–6].

The hemp cultivation ensures positive impact on preserving biodiversity, improvement of soil quality, and dredging of heavy metals from the ground and, from the other side, supports cleaning of air by the absorption of CO_2 from atmosphere. Hemp is a yearling fast growing plant with well-developed leaves system, which results in that one season of hemp cultivation causes absorption about 10 tons of CO_2 from the air. Hemp is resistant to drought; due to a long-root system, dense hemp cultivation causes effective blocking of weed growing, which means significantly less consumption of water and pesticides in comparison to cotton. The environmental approach to flax/linen processes based on Life cycle assessment (LCA) study for linen shirt production confirms that the impacts of the linen shirt are up to seven times smaller than the impacts of the cotton shirt in terms of the most relevant environmental indicators, such as the freshwater aquatic ecotoxicity potential or water consumption. In the case of the global warming potential or the primary energy consumption, the environmental impacts of the linen shirt production are either equivalent to those of the cotton shirt or 10–15% higher [7].

Reports on linen/hemp textile bioactivity and its effect on human in terms of the well-being and health of wearer are very limited in the available literature.

The goal of this chapter is to provide knowledge covering a new approach to flax/hemp fibers regarding their potential to give new properties of clothing, which are able to ensure optimal environment for human body and improve health and quality of life. The properties of the textiles are determined by carefully selected fibrous plant variety and fiber processing suitable to final application.

This chapter covers discussion about new approach to the flax and hemp fibers taking into account their features and potential to have a positive impact on human life, based on authors' finding and other available literature.

The authors introduce multiperspective meaning of the term "improvement of human life" to demonstrate environmental and human-ecological performance of flax/hemp fibers.

The discussed textile plant raw materials have positive impact on the following:

1. environment—ensuring contribution to striving toward keeping green planet and healthy life for current and future generation,

- hemp and flax cultivation protects agriculture areas against loss of biodiversity, which is important for limitation of greenhouse gasses emission,
- hemp improves the productivity of the soil, removes heavy metals, and can be used for soil remediation and reclamation at industrial area,
- 1 ha of hemp plantation absorbs about 10 tons of CO_2 from atmosphere every year,
- hemp and flax have potential to cascade the use of the whole biomass delivering war materials to different sectors of economy, according to strategy "zero waste,"
- fibrous plants deliver raw materials for production of bioproducts giving possibility to replace nonrenewable raw materials, e.g., construction, plastics, textiles, with natural ones
- hemp and flax fibers are renewable and biodegradable;
- the pursuit of plastic elimination in many sectors, including everyday life contribute to the reduction of greenhouse gasses emissions and, consequently, decarbonization,

2. multifaced human-ecological performance—linen and hemp textiles based on their inherent properties ensure comfort, well-being, and health-supporting properties for users target improving of everyday life.

Both the aspects of environmental and human-ecological performance of flax and hemp have to be identified as complex factors, which have a significant effect on the improvement of human life.

2. Characteristic of bast fibers: structure and chemical composition

Flax and hemp are the most popular bast fibers, which can be used for textile purpose, including apparels. The lignocellulosic fibers are delivered by yearling plants with potential of multiple applications.

The structure of stem of flax and hemp is very similar; fibers are created as concentric rings around lumen and xylem in the whole length of stem parallel to the stem axis. **Figure 1** shows the cross section of flax and hemp stem [8, 9].

The cellulosic structure inside the secondary cell wall of fiber is schematically presented in **Figure 2**.

Flax and hemp fibers occur in form of bundle called technical fibers containing elementary fibers glued by pectin and lignin as well as naturally connected together due to their arborescent structure (**Figure 3**).

2.1 Chemical composition

Bast fibers contain cellulose, lignin, hemicellulose, pectin, waxes, and fats in their chemical composition. The schematic image of the chemical components distribution in

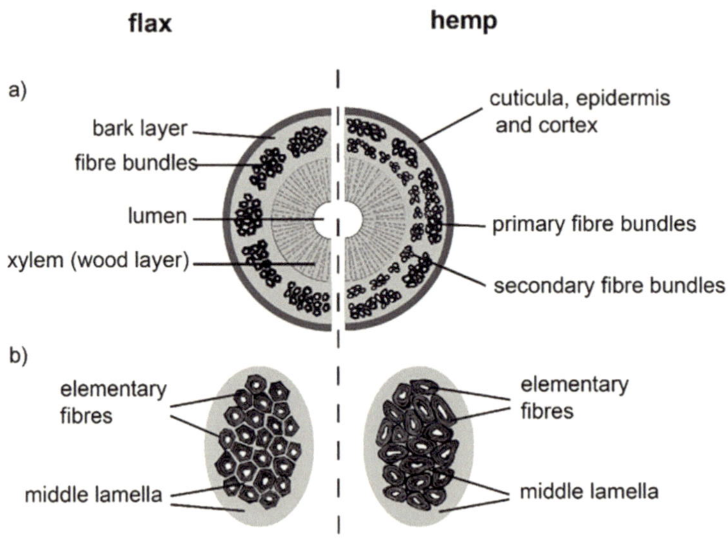

Figure 1.
Structure of flax and hemp stem. (a) Cross section of stem. (b) Fiber bundle.

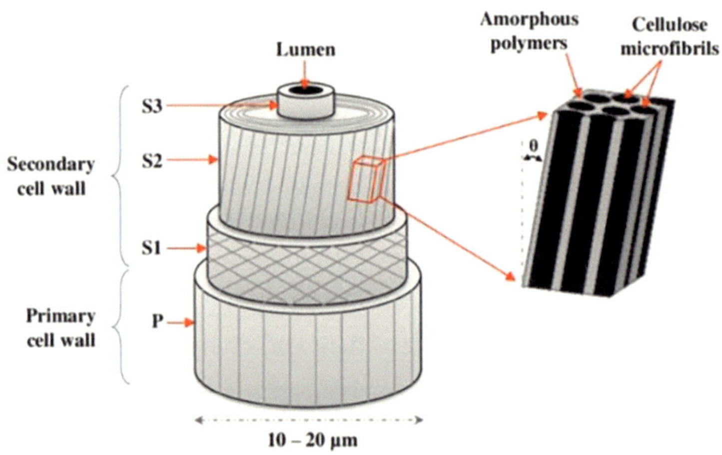

Figure 2.
Schematic depiction of the microscopic structure of an elementary flax fiber showing the cellulosic structure inside the secondary cell wall (figure redrawn from [10]).

the cell wall of the fiber is presented in **Figure 4**. The figure illustrates schematically in which way the chemical components of bast fibers are distributed in the fiber structure.

The share of the chemical components depends on the fibrous plant variety and the applied method of fiber extraction. From this reason, different values of the cellulose, lignin, pectin, and hemicellulose waxes are given by different authors in their scientific articles [14–18].

The diversification of fiber chemical composition [19] resulting from fibrous plan variety, applied method of retting, fiber extraction, and subsequent stages of processes is presented in **Table 1**.

Figure 3.
Real and graphical image of naturally arborescent structure of flax/hemp fibers (based on [11, 12]).

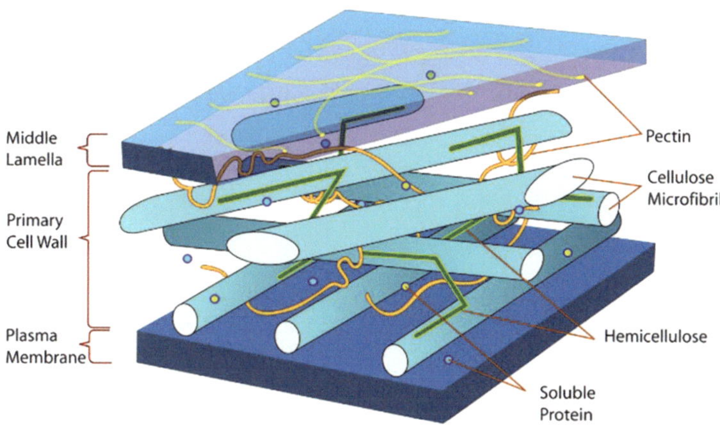

Figure 4.
Schematic image of the section of a hemp cell wall [13].

The cultivar of fibrous plants, the method of plant growing, and the method of fiber extraction have to be selected taking into account the obtaining of the fiber with chemical composition suitable for textile purpose. The high cellulose content gives softness to the fiber and ensures efficient spinnability in opposite to lignin and pectin. Their big share in the fiber results in low fiber quality, e.g., high stiffness and high linear density coming from inefficient fiber separation, which makes the spin process difficult.

3. Bast fiber processes

Flax/hemp fiber extraction from the stem is strongly connected to the retting process, which is usually a microbiological process that uses bacteria and fungi strains applied in order to degrade pectin and lignin, remove woody part of the stalk, and divide the technical bundles to smaller fiber complexes. There are several retting methods: water, dew, chemical, enzymatic, and physical retting. The use of the three

Degumming method	Variety	Content of:									
		Waxes and fats		Pectin		Lignin		Celullose		Hemicelullose	
		%	SD	%	SD	%	SD	%	SD	%	SD
HEMP											
Water retting	Beniko	0.23	0.01	1.47	0.09	2.81	0.29	71.31	1.32	15.03	0.02
	Wojko	0.24	0.04	0.67	0.02	3.02	0.31	72.53	0.11	16.67	0.24
	Tygra	0.25	0.04	0.56	0.00	2.78	0.28	70.79	0.13	15.00	0.28
	Białobrzeskie	0.34	0.02	0.67	0.02	2.38	0.22	72.03	0.22	14.37	0.29
Decortication		0.47	0.02	2.00	0.09	5.55	0.17	66.02	0.46	21.25	0.05
Dew retting		0.56	0.14	3.68	0.19	4.31	0.04	66.16	0.48	21.72	0.12
Water retting	Białobrzeskie	0.34	0.02	0.67	0.02	2.38	0.22	72.03	0.22	14.37	0.29
Osmotic degumming		0.44	0.04	2.82	0.22	4.03	0.09	67.81	0.52	16.29	0.03
FLAX											
Decortication		1.26	0.00	4.62	0.16	4.00	0.16	68.89	1.91	29.35	0.16
Wet degumming +ultrasound	Modran	0.69	0.07	4.41	0.50	4.20	0.16	75.54	1.18	19.62	0.15
Cottonization		0.97	0.10	4.72	0.39	4.26	0.15	73.51	0.98	16.44	0.23
Decortication		1.47	0.07	4.11	0.38	8.60	0.30	64.57	0.85	29.38	0.08
Wet degumming +ultrasound	NIKE	0.76	0.00	3.56	0.27	4.46	0.48	77.44	1.58	16.43	0.25
Cottonization		0.95	0.05	2.39	0.22	4.87	0.51	74.25	0.20	13.84	0.06
Decortication		1.47	0.07	4.11	0.38	8.60	0.30	64.57	0.85	29.38	0.08
Wet degumming +ultrasound	B14 Iung	1.33	0.01	5.43	0.28	6.69	0.48	75.04	0.46	23.92	0.02
Cottonization		1.72	0.09	3.57	0.23	6.10	0.01	72.20	0.47	20.41	0.09

Table 1.
Chemical composition of fibers coming from different varieties of flax controlled in sequence stages of fiber processing; different varieties of hemp extracted with use of dew retting as well as hemp fiber Białobrzeskie variety obtained by different methods of fiber extraction: Decortication, osmotically degumming, and water retting [19].

last listed methods is limited to small application; these methods have not been used in large industrial scale yet.

The traditional water retting process delivers the best quality of long fibers characterizing by high cellulose content and low lignin and pectin share resulting in low fiber linear density, good mechanical properties, and good spinnability. Water retting used to be the most recommended process for bast fiber dedicated to textile production. Nevertheless, owing to the generation of large amounts of wastewater and high pollution of soil and air, this method has been abandoned in Europe and many other countries in the 1960s [20–22].

The second retting method delivering good quality of long fibers is dew retting; nevertheless, the process effectiveness and quality of the fibers are not stabile because it depends on weather conditions prevailing in time of the stalk remaining on the field.

The method of fiber extraction, which allows one to avoid the retting process, is the decortication of raw straw collected when the plants reach proper fiber maturity. Decortication is a mechanical process giving one type of insufficiently dividing fibers with high impurities content.

The method of flax/hemp fiber extraction determines further technological chain of fiber processing and spinning. The simplified schema of value chains of bast-fiber-based textile for both the methods of fiber extraction, e.g., with the use of retting and with application of decortication, is presented in **Figure 5**.

Flax/hemp stalks after retting are mechanically breaking and scutching to separate fibers from the woody part of the plant and dividing big complexes of fibers to smaller ones. The processes are efficient due to the decomposition of noncellulose components made by bacteria and fungi activity during retting. As a result of scutching, long fibers and scutched tow are produced. Long fibers must be mechanically cleaned and straighten as well as laid parallelly toward each other with the use of hackling machine dedicated only to bast fibers processing. The hackling process delivers sliver of long fibers and mass of short fibers, e.g., hackling tow. Both types of fibers are spun with application of linen spinning system, long fibers with the use of wet spinning, and short fibers by wet or dry spinning.

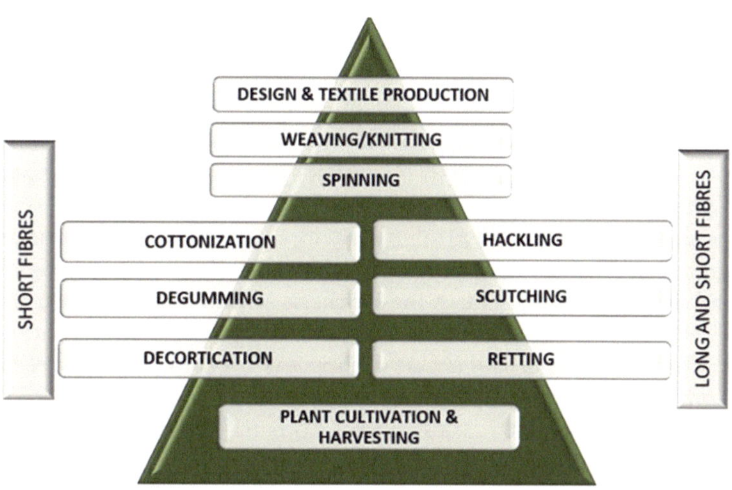

Figure 5.
Simplified value chain of hemp/flax textile product [own elaboration].

Second value chain determined for decorticated fibers allowing avoid of retting has to employ the wet degummed process to remove lignin, pectin, and partially hemicellulose in order to make it possible to divide the bast fibers for elementary ones. Using this system, only one type of fibers is produced; it is not possible to obtain long and separately short fibers. Degummed decorticated fibers are cottonized to make the fibers suitable for spinning with the use of the cotton spinning system [23].

The yarn prepared by using the cotton spinning system, including decortication, and the linen spinning system can be used for textile purpose.

4. Bioactivity of bast fibers

4.1 Antioxidant activity

Flax and hemp fibers apart from the main components as cellulose, hemicellulose, pectin, lignin, fats, and waxes contain in their chemical composition phenolic acids, which are natural antioxidants [24]. Based on an effectively scavenging chain reaction and deleterious radicals as well as suppressing radiation-induced oxidative reactions, phenolic acids serve for preserving the physiological integrity of plant cells exposed to both air and impinging ultraviolet (UV) radiation [25].

The study described in [19] confirmed the diversified presence of ferulic, *p*-coumaric, syringic acids and small amounts of sinapinic acid in the chemical composition of flax and hemp fibers. The differences resulted from various fibrous plant varieties and the method of fiber extraction, which have effect on their chemical composition. Syringic acid naturally occurring in O-methylated phenolic acid shows high antioxidant and antibacterial activity and can be enzymatically degraded by some bacteria as a source of methane or methanol [26]. As is visible in **Table 2**, the decorticated fibers of hemp Bialobrzeskie as well as Modran and B14 flax varieties show the highest content of syringic acid. The lower share in retted fibers results from the fact that syringic acid can be degraded during the biological retting process.

Sinapinic acid (3,5-dimethoxy-4-hydroxycinnamic acid) exhibits antioxidant, anti-inflammatory, anticancer, antimutagenic, antiglycemic, neuroprotective, and antibacterial activities [27]. Sinapinic acid occurs only in hemp fibers; the highest amount was detected in the decorticated Bialobrzeskie hemp, as well as in the water-retted Wojko and Tygra hemp.

Coumaric and ferulic acids are the main hydroxycinnamic acids in flax [28]. *p*-coumaric acid is mainly a plant metabolite, which exhibits antioxidant and anti-inflammatory properties. It also shows bactericidal activity by damaging bacterial cell membrane and by interacting with bacterial DNA.

Ferulic acid, together with dihydroferulic acid, is a component of lignocellulose, serving to crosslink the lignin and polysaccharides, thereby conferring rigidity to the cell walls [29]. It is an intermediate in the synthesis of monolignols, the monomers of lignin, and is also used for the synthesis of lignans. Ferulic acid shows antioxidant and anti-inflammatory properties and is able to suppress UV-radiation-induced oxidative reductions, which has a negative effect on the skin.

Ferulic acid is easily soluble in water and can be easily removed from the fibers during the retting process; however, coumaric acid is poorly soluble in water, and its removal could be only partial. Among all types of tested flax and hemp fibers, almost all the decorticated fibers contained the highest amount of *p*-coumaric and ferulic acids because decortication is an entirely dry mechanical process.

Degumming method	Variety	Content of acids:							
		Syringic [mg/100 g]		Sinapinic [mg/100 g]		p-coumaric [mg/100 g]		Ferulic [mg/100 g]	
		Result	±SD	Result	±SD	Result	±SD	Result	±SD
HEMP									
Water retting	Beniko	0.031[a]	0.001	-*[a]	—	0.722[a,b]	0.019	-*[a]	—
	Wojko	0.046	0.001	0.048	0.002	0.741[a]	0.006	0.027[a]	0.001
	Tygra	0.036	0.001	0.100	0.003	0.695[b]	0.034	0.572	0.031
	Białobrzeskie	0.033[a]	0.001	-*[a]	—	0.024	0.006	-*[a]	—
Decortication		0.224	0.011	0.672	0.023	0.746	0.008	2.082	0.036
Dew retting		0.079	0.003	-*[a]	—	0.717	0.008	0.039[a]	0.004
Water retting	Białobrzeskie	0.033	0.001	-*a	—	0.024	0.006	-*[a]	—
Osmotic degumming		0.094	0.003	-*[a]	—	1.111	0.011	0.625	0.009
FLAX									
Decortication		0.235	0.008	-*[a]	—	0.995	0.024	5.749	0.159
Wet degumming + ultrasound	Modran	-*	—	-*[a]	—	-*[a]	—	0.041	0.002
Cottonization		0.035	0.002	-*[a]	—	-*[a]	—	1.206	0.053
Decortication		-*[a]	—	-*[a]	—	-*[a]	—	1.485	0.034
Wet degumming + ultrasound	NIKE	-*[a]	—	-*	—	-*[a]	—	0.054	0.001
Cottonization		-*[a]	—	-*[a]	—	-*[a]	—	1.035	0.020
Decortication		0.125	0.060	-*[a]	—	0.904	0.009	3.146	0.106
Wet degumming + ultrasound	B14 IUNG	0.052[a]	0.002	-*[a]	—	0.027	0.008	2.525	0.106
Cottonization		0.040[a]	0.001	-*[a]	—	0.756	0.034	1.736	0.045

- not identified.*
a and b represent the groups for which the mean values do not differ statistically at the assumed significance level. The mean values labeled with the same letter (a or b) do not differ statistically at (α = 0.05).

Table 2.
Acid content in the flax and hemp fiber. Results are expressed as mean ± standard deviation (SD), $n = 4$. Lowercase letters indicate significant differences at $p \leq 0.05$ according to the Tukay's HSD test [19].

Because the flax and hemp fibers contain phenolic acids in their chemical composition, they show antioxidant activity, which is presented in **Table 3**. The fiber antioxidant activity depends on the type and variety of fibrous plant as well as on the method of fiber extraction [30]. The highest value of ferric reducing antioxidant power (FRAP) and value of 2,2,2-diphenyl-1-picrylhydrazyl (DPPH radical) scavenging activity illustrating antioxidant activity of the fibers were found in all types of fibers after decortication because the fibers contain the largest amount of phenolic acids.

4.2 Antibacterial activity

The tests of antibacterial properties of flax fibers were conducted with a fluorescent method using strains of the clinical bacteria *Staphylococcus aureus* isolated from ill people [31].

Degumming method	Variety	FRAP [μmol/L]		Inhibition of DPPH [%]	
		Result	±SD	Result	±SD
HEMP					
Water retting	Beniko	140.34	4.75	11.30[a]	0.92
	Wojko	156.75	2.31	10.04[a]	0.49
	Tygra	165.76	1.62	32.55	0.32
	Białobrzeskie	76.62	1.33	3.09	0.18
Decortication		230.22	1.55	18.03	0.63
Dew retting		124.09	1.93	5.31	0.25
Water retting	Białobrzeskie	76.62	1.33	3.09[a]	0.18
Osmotic degumming		93.71	0.69	3.94[a]	0.19
FLAX					
Decortication		523.00	2.08	33.85	0.17
Wet degumming + ultrasound	Modran	129.45	1.28	5.80	0.14
Cottonization		70.69	2.31	3.29	0.24
Decortication		519.75	2.69	29.76	0.15
Wet degumming + ultrasound	NIKE	140.79	1.16	7.64	0.16
Cottonization		78.84	1.59	5.10	0.36
Decortication		485.84	2.11	37.71	0.14
Wet degumming + ultrasound	B14 Iung	195.79	2.23	11.05	0.11
Cottonization		119.42	4.55	6.82	0.23

a represents the group for which the mean values do not differ statistically at the assumed significance level. The mean values labeled with the same letter (a) do not differ statistically at (α = 0,05).

Table 3.
The antioxidative activity of different varieties of bast fibers depending on the extraction method. Results are expressed as mean wf standard deviation (SD), n = 3. Lowercase letters indicate significant differences at $p \leq 0.05$ according to the Tukay's HSD test [19].

Test conditions:

- Bacterial suspension density: 0.5 McF,
- Incubation conditions: temperature—37°C, time—90 min.

S. aureus belongs to the most common etiological factors of inflammation in surgical places. The evaluation of bacteria viability was performed using the Bacterial Viability Kit (Molecular Probes).

The images presented in **Figures 6** and 7 show the fibers and strains of the clinical bacteria *S. aureus* isolated from ill people after 90 minutes of incubation. The orange particles on the fiber surface illustrate killed bacteria. It is visible that killed bacteria are located only in direct contact with the fibers.

The study [31] on the antibacterial activity of flax covered five varieties: Artemida, Modran, Sara Nike, and Luna. The extracted fibers with use of dew retting or water retting method were tested in order to evaluate their ability to reduce colonies of *S. aureus bacteria*. The results confirmed that the fiber antibacterial capacity is strongly related to plant variety and the applied extraction method, which have effect on fibers chemical composition, mainly content of lignin cross-linked with phenolic compounds. Percentage differences in the antibacterial capacity of all types of flax fibers are shown in **Figure 8**.

Dew-retted flax fibers coming from all tested plant varieties showed stronger capacity to reduce *S. aureus* bacterial colonies than water-retted fibers. This resulted from the fact that some types of phenolic acids, for example, ferulic acid, are water soluble and have been removed during the water-retting process.

4.3 Antimicrobial modification of bast fibers

4.3.1 Genetic engineering

In order to improve antibacterial properties of flax fibers, researchers applied genetic engineering to modify plant DNA [32]. Transgenic flax plants overproducing

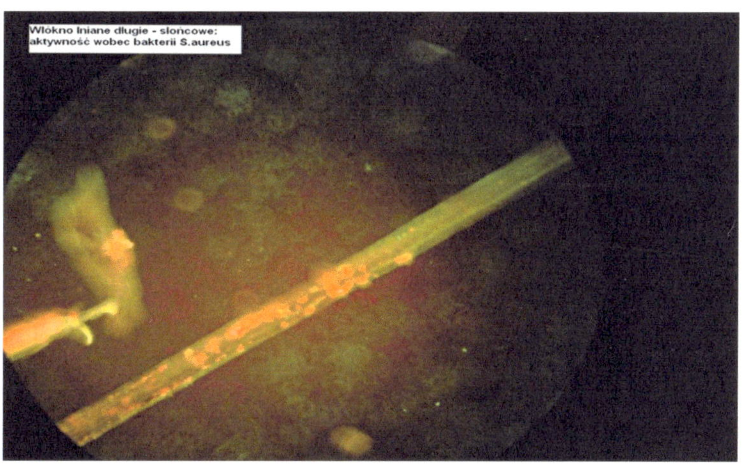

Figure 6.
Dew retted flax fibers activity toward staphylococcus aureus bacteria.

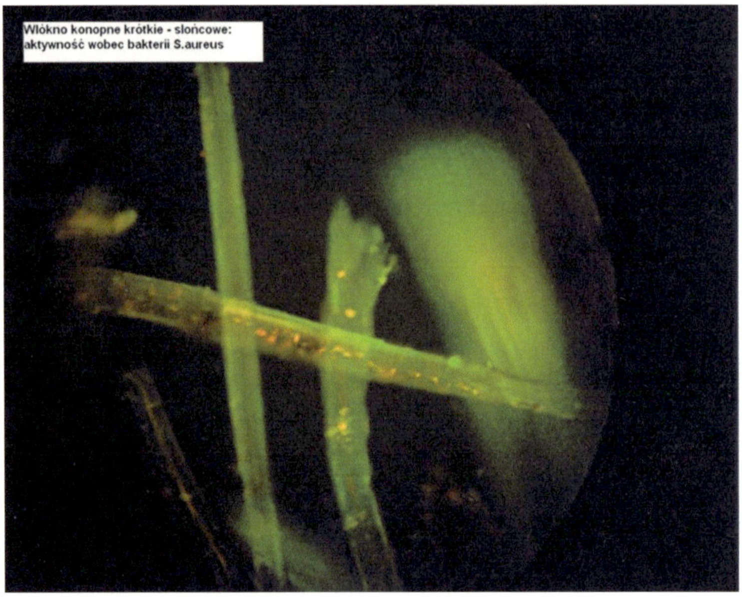

Figure 7.
Dew retted hemp fibers activity toward staphylococcus aureus bacteria.

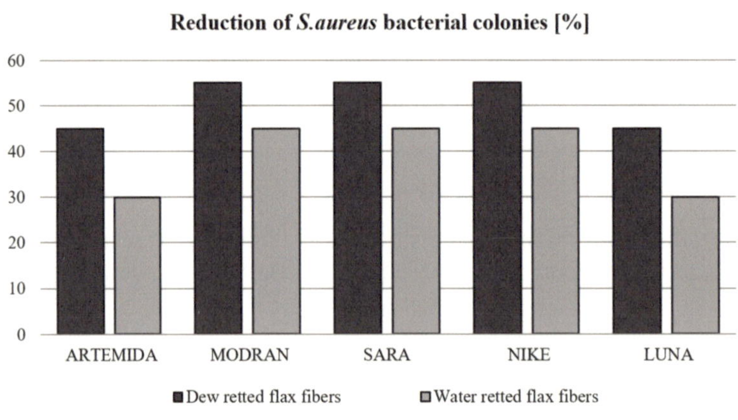

Figure 8.
Comparison of antibacterial capacity of flax fibers.

compounds from phenylpropanoid pathway accumulate phenolic derivatives of potential antioxidative and, thus, antimicrobial activity. The researchers showed that the extract alkali hydrolyzed seedcake had antibacterial activity, which might be useful as a prophylactic against bacterial infection.

Although the GMO plant modification is not allowed in Europe due to the strategy "Europe free from GMO," studies on fibrous plant genetic engineering are very limited.

4.3.2 Chemical modification

The way to obtain stronger antibacterial properties of bast fibers is the chemical modification of linear or flat textiles. Flax and hemp as cellulosic fibers are characterized with high reactivity due to containing hydroxy group in compounds in their chemical composition.

Racu conducted study on the application of grafting of monochlorotriazinyl-β-cyclodextrin on hemp fiber stream at the time of wet spinning. Four compounds, e.g., ferulic acid, caffeic acid, ethyl ferulate, and allantoin, have been included into the cavities of monochlorotriazinyl-β-cyclodextrin and grafted on hemp fibers. Obtained yarn showed that the method allowed for a significant modification of Sano Genetics properties of the hemp fibers [33].

The application of silver in shape of nanoparticles is the most common method to functionalize different textile materials, including natural fibers. Research on the functionalization of the scoured flax fibers by the insertion of silver nanoparticles was conducted with the use of two different methods: first where Ag + was reduced by using the functional groups of flax in the internal reduction, and second where trisodium citrate was used as an external reducing agent in the external reduction method [34]. The modified scoured flax fibers with silver nanoparticles showed very good barrier properties against UV radiation and excellent antibacterial activity.

A good example of nanosilver application for hemp fiber to make strong antibacterial activity is the use of selective 2,2,6,6-tetramethylpiperidine-1-oxy radical (TEMPO)-mediated oxidation, i.e., oxidation with sodium hypochlorite, catalytic amount of sodium bromide, and the 2,2,6,6-tetramethylpiperidine-1-oxy radical (TEMPO), followed by silver sorption from aqueous silver nitrate solution, described by Milanović et al. [35]. The introduced hydrophilic carboxyl in hemp fibers caused by TEMPO-mediated oxidation influenced increased silver sorption and, in consequence, gave efficient antibacterial activity. The TEMPO-oxidized hemp fibers with absorbed silver showed good antibacterial activity against the tested bacterium strains: *S. aureus* and *Escherichia coli*, and the fungus *Candida. albicans*.

The new method of improving of antibacterial activity of hemp fibers has been developed by Chang [36]. The research group developed method of grafting of hemp fiber with the use of quaternary ammonium groups (HF–GTA) prepared by alkalization, oxidation, amination, and quaternization multistage reactions, whereby the grafting reaction mainly takes place on the cellulose and hemicellulose hydroxyl groups, without negative effect on fibrous morphology, thermal stability, and hygroscopicity. This method gives good antibacterial activity of hemp fiber against bacteria strains *E. coli* and *S. aureus*. The obtained barrier properties are characterized by good washing resistance.

5. Well-being and comfort

Physiological comfort determined by skin parameters, such as temperature and moisture, is affected by raw materials as well as clothing design and depends particularly on the suitability of worn type of clothing to the level of physical activity of user and ambient climatic conditions.

Hemp and flax fibers show high ability to moisture sorption from ambient air. The hygroscopicity of lignocellulosic fibers depends on the relative humidity of air, which is shown in **Table 4**. The fibers can absorb more moisture in conditions of higher relative humidity of air, which means, in practice, that the linen/hemp clothing easily absorbs sweat produced by the human body. In the case of conditions of everyday life and moderate physical effort, the clothing ensures optimal comfort to the wearer, allowing easy skin breathing and air exchange from the

Fiber	Moisture content in condition of different relative humidity of air [%]					
	30	40	50	60	70	100
Flax	7.5	8.3	9.1	9.9	10.7	23
Hemp	8.0	8.7	9.4	10.1	10.8	24

Table 4.
Effect of relative air humidity on the hygroscopicity of selected bast fibers [17].

area of skin clothing to outside. In the case of high physical effort, doing sports, when the intensity of sweating is high, the T-shirt made of cellulosic raw material becomes uncomfortable due to feeling of wet touch. Linen/hemp clothing should be dedicated to everyday life, indoor and outdoor work, leisure, tourism, and other activities with low and moderate effort, when the garment ensures optimal comfort to the human body [37].

The comparative study on comfort parameters of linen and linen/polyethersulfone (PES) clothing during moderate physical exercise was described by Zimniewska [38]. Tested shirts and trousers were prepared from linen and polyester fiber in different composition of the raw materials; share of linen in the blend with PES increased by 25% in each following clothing sample. The fabrics were characterized by the similar density of threads; linen and PES yarns produced to this experiment have similar linear density.

Values of the parameters that affected the comfort of clothing are presented in **Figures 9–12**.

The fabric hygroscopicity increased with increasing of the share of linen in the blends linen/PES; similarly, the time needed to drop water sorption is longer in the case of bigger PES content in the blend. The linen fabric showed the biggest value of air permeability; the lowest one was observed for 100% PES fabric. The evaluation of moisture of back skin of clothing users during 5 hours of experiment was conducted in order to assess the comfort. During the experiment, volunteers did not perform any physical activity. Results of the experiment are presented in **Figure 13**.

The results proved that the skin moisture was the lowest in case of wearing the clothing made of 100% linen; however, the biggest increasing of moisture was observed when PES clothing was worn.

Bast fiber does not gather electrostatic charges on their surface, which results in their high ability of moisture sorption. The values of electrostatic potential of the tested linen/PES fabrics are presented in **Figure 14**, where the highest value observed for PES fabric decreased when linen share in the blends increased. The presence of electrostatic charges on the synthetic clothing surface results in some kind of skin and mental irritation of garment users.

6. Positive effect on the human body

Clothing covering the body of the user in the conditions of daily life affects their physiological parameters. The clothing made of bast fibers or made from blend bast fibers/synthetic fibers with high share of natural fibers provides the optimal comfort for users and affects some physiological parameters of the human body.

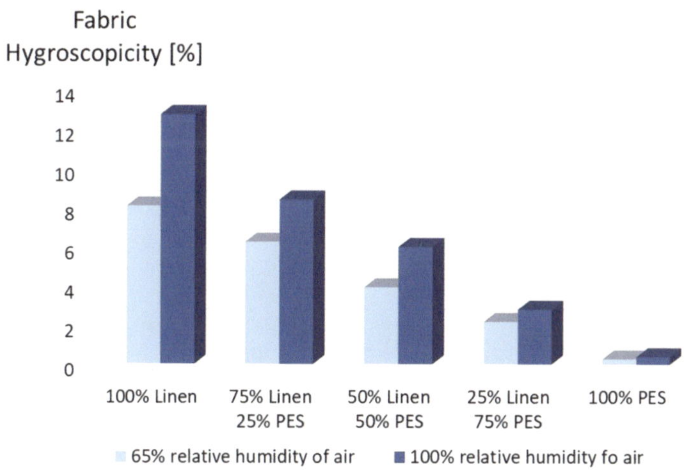

Figure 9.
Hygroscopicity of linen/PES fabrics tested in 65% and 100% relative humidity of air (based on [38]).

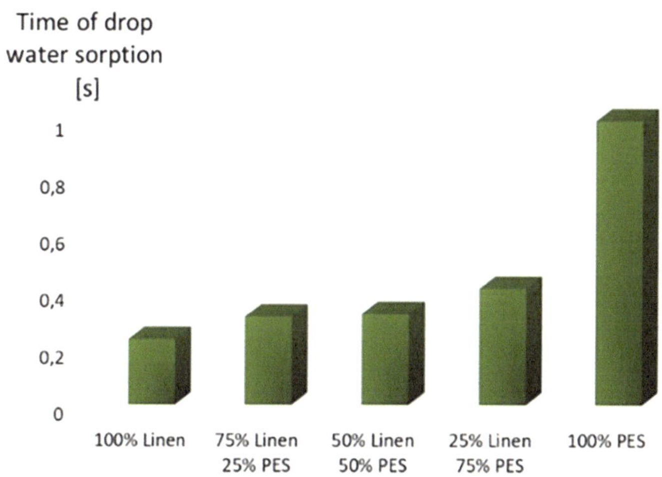

Figure 10.
Time needed to drop water sorption by tested linen/PES fabrics (based on [38]).

6.1 Muscle tension and fatigue tendency

The human skin parameters that determine the comfort during clothing wearing in ambient climatic conditions and given level of physical activity are strongly related to well-being and can affect tendency to tiredness [38, 39].

The study on the monitoring of shoulder muscle tension of volunteers wearing linen/PES clothing with different shares of both components allowed identifying the phenomenon of clothing influence on electromyographic parameters of the muscles [38]. Examples of global electromyographic records taken at resting state from a volunteer wearing 100% linen clothing (a), 100% linen clothing double layers (b), 50% linen 50% PES clothing (c), and 100% PES are presented in **Figure 15**. The results of the tests proved that polyester clothing can change

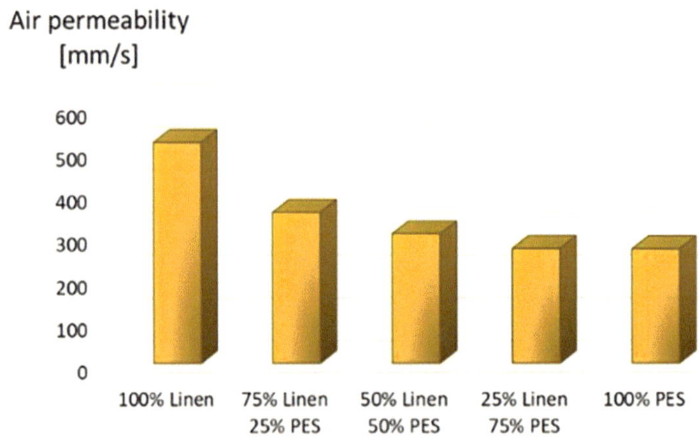

Figure 11.
Air permeability of linen/PES fabrics with different share of both types fiber (based on [38]).

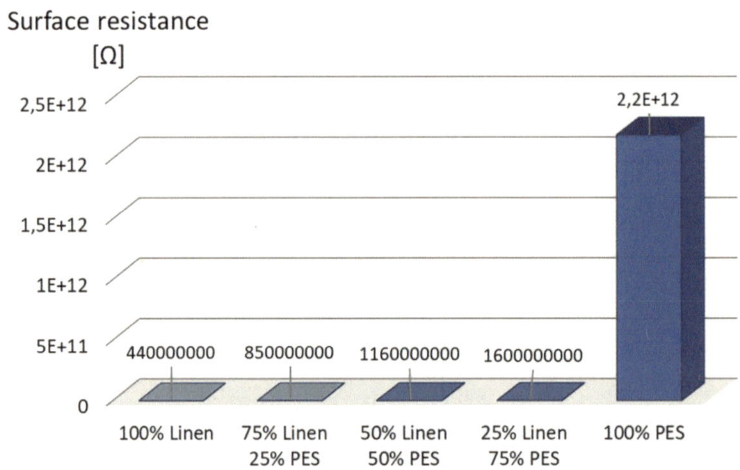

Figure 12.
Surface resistance of linen/PES fabrics with different share of both types fiber (based on [38]).

muscle electromyography (EMG) records in the range of amplitude and frequency after 5 hours of covering on the body with this kind of apparel, which indicates the occurrence of desynchronization of motor units. In the case of wearing linen clothing, such a phenomenon did not occur. The PES clothing gathered electrostatic charges on its surface causing increase in the mean value of frequency of motor units in the resting state. More intensive sweating of volunteers' body under the influence of PES clothing caused changes in the EMG amplitude in the resting state and amplitude and frequency during voluntary movement. The threshold value of polyester fiber share in blend with linen fiber is 25%. The clothing made of 100% linen fabric or 75% linen/25% PES did not cause the desynchronization of motor units in healthy muscles and provided optimal comfort of apparel use. Changes in the activity of the motor units of muscles found during the study proved that the

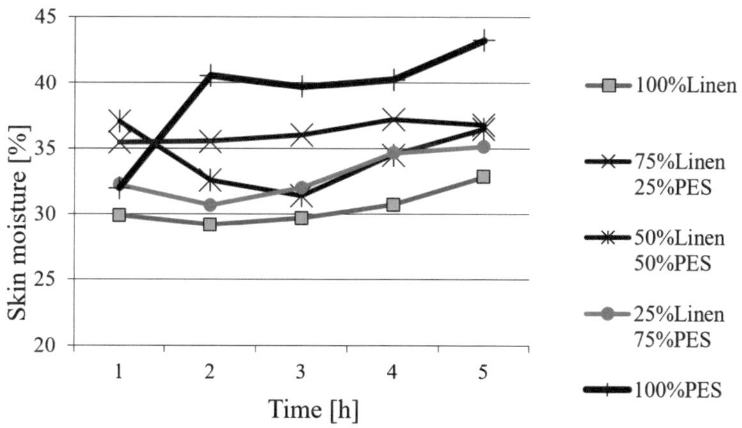

Figure 13.
The moisture of back skin under tested clothes during 5 hours of experiment (Grant 1000 series squirrel – Operating instruction) [38].

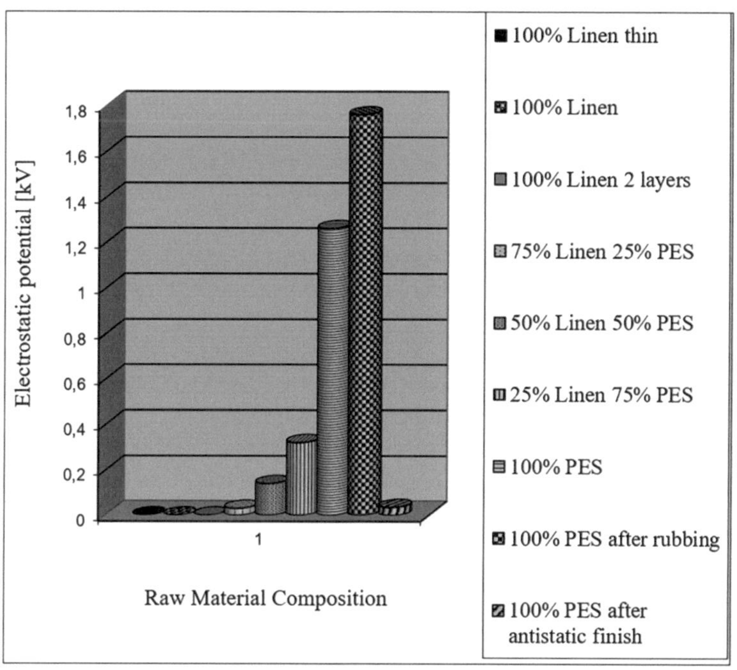

Figure 14.
The electrostatic potential on the surface of tested fabrics [39].

polyester clothing can cause increase of tendency toward general fatigue of people wearing the clothing.

6.2 Oxidative stress

Study on clothing influence on some parameters of human physiology covered the experiment on oxidative stress affected by clothing made of different raw materials

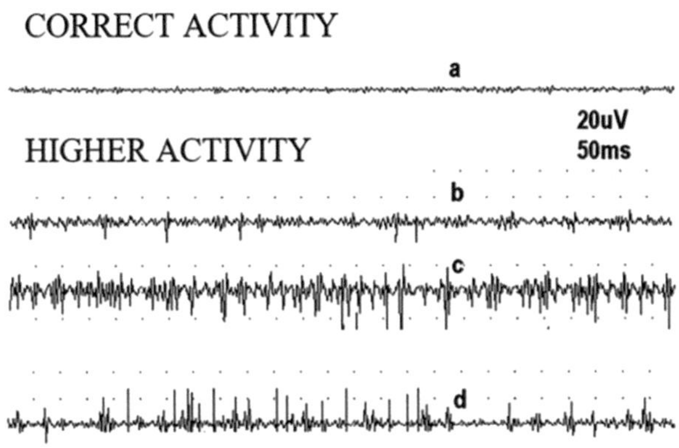

Figure 15.
Examples of global electromyographic records taken at resting state from a volunteer wearing: 100% linen clothing (a), 100% linen clothing double layers (b), 50% linen 50% PES clothing (c), and 100% PES (d) [39].

[37]. The experiment was done with volunteers wearing linen and then PES clothing in given climatic conditions during an 8-hour rest period, 20-minute moderate physical activity at a level of 75 W, on a stationary bike and the period of returning to baseline. One of the ways to test the ability of the organism to defense itself against the reactive oxygen species is to determine the so-called total antioxidant status (TAS). This parameter informs about the total ability of tissues to neutralize exactly determined amount of reactive oxygen species. The TAS was tested based on blood analysis after each stage of the volunteers' activity.

The results of the TAS test are shown in **Figure 16**. The lower level of total antioxidative status in individuals wearing polyester clothing indicated that, probably, this is an effect of increased production of reactive oxygen species, which are responsible for the oxidative stress. The phenomena observed during conducted tests confirmed the stipulations that the wear made of linen—a natural fiber—not only guarantees a comfort, but may also have a positive effect on users' health. The clothes made from polyester fibers can have an unfavorable effect on human organism.

These TAS changes tested in conditions of increased sweating, i.e., during exercise, were statistically significant. The results of the research resulted from the following phenomena accompanying of polyester clothing:

- accumulation of electrostatic charges on the surface,
- low permeability,
- low hygroscopicity,
- cause of sweating increase,
- cause increase of body temperature.

6.3 UV barrier

Garment barrier properties against ultraviolet radiation are mainly determined by the structure and thickness of fabric or knitted fabric used for clothing preparation. High fabric density and minimalized porosity play the most important role to block

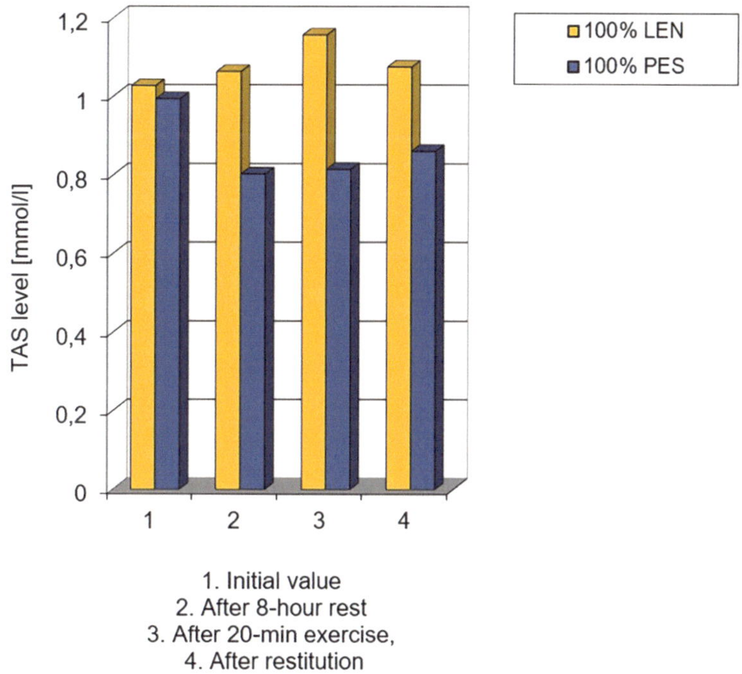

Figure 16.
The values of total antioxidant status (TAS) in volunteers wearing linen and PES clothing [40].

UV rays transitioning through the fabric and make it impossible to touch the skin of the wearer.

Even in the dense structure of fabrics and proper thickness to make mechanical barrier against radiation, the clothing does not always ensure safety for wearers in conditions of intensive solar radiation.

Bast fibers contain in their chemical composition lignin and phenolic acids, including ferulic acid, with ability to absorb ultraviolet rays. The values of ultraviolet protection factor (UPF) of tested samples of linen fabrics characterized by different mesh/hollows content in their structure are shown in **Figure 17** [41]. Clothing made of bast fibers with proper parameters of fabric structure offers very good UV barrier properties—guarantee safety and optimal comfort for users under conditions of solar radiation.

In case of using the garment in extremely high UV index area, the inherent protective properties of bast fibers can be supported by the application of UV absorbers, for example, by nanolignin coating, which is the most ecological method of fabric modification [42, 43].

7. Bioactive bast fiber textile products

7.1 Clothing as a supplement of skin disease treatment

The bast fibers showing inherent antibacterial and antioxidant activity and properties guaranteeing optimal comfort of clothing are suitable raw materials dedicated to

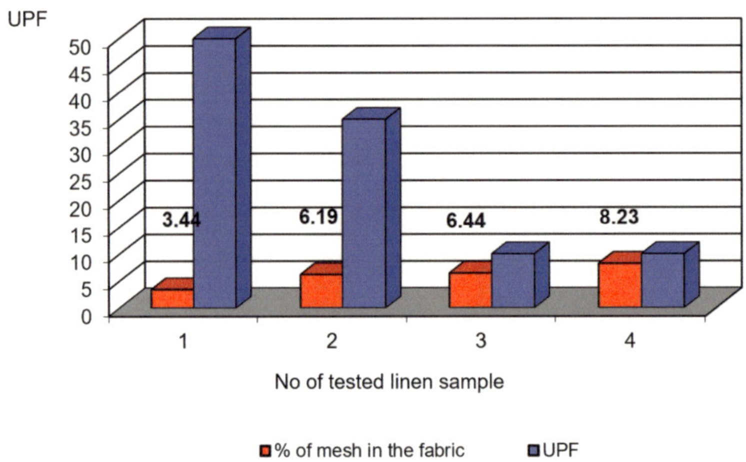

Figure 17.
"Mesh" size and UPF of 100% flax fabrics [41].

The extract	Flavonoids expressed as quercetin [%]	Polyphenols expressed as rosemarinic acid [%]	Tannins expressed as pyrogallol [%]
Green tea	1.01 ± 0.01	5.63 ± 0.38	15.14 ± 0.15
Viola tricolor	7.61 ± 0.06*	0.44 ± 0.03	0.72 ± 0.12

Table 5.
The content of polyphenolic compounds in ethanol-water extracts (1:1) from the tested raw materials.

functional textile manufacture. One of the examples of flax fiber functional textile is clothing, which acts as a supplement of dressing addressed the treatment of dermatological diseases [44]. The clothing was made of combination of two raw materials: linen for active part of clothing and organic cotton for other clothing elements. From both the raw materials, only linen-knitted fabric was enriched with biologically active medicinal plants extracts closed in microcapsules to ensure high bioactive performance of the selected clothing parts. Microcapsules containing herbal extracts are fixed to the inner clothing layer in order to guarantee direct and continuous contact of herbs with ill skin. The fabrics were dyed with natural herbal extracts to avoid allergic reaction and guarantee safety for dermatological patients during clothing wearing. To ensure an effective treatment of skin problems, herbal extracts, viola tricolor and green tea with proved properties suitable for dermatitis healing, were implemented to this solution. The content of active components in herbal extracts is presented in **Table 5** [45].

The results of tests conducted during 5 weeks of everyday functional clothing wearing by dermatological patients confirmed efficiency of applied method of skin treatment in improving of skin condition determined by skin moisture content (**Figure 18**) and transepidermal water loss (**Figure 19**) as well as lowering of skin illness sensation, e.g., itching feeling (**Figure 20**).

The observed changes of skin parameters tested before and after the experiment of clothing wearing proved that clothing made of naturally dyed linen-knitted fabric

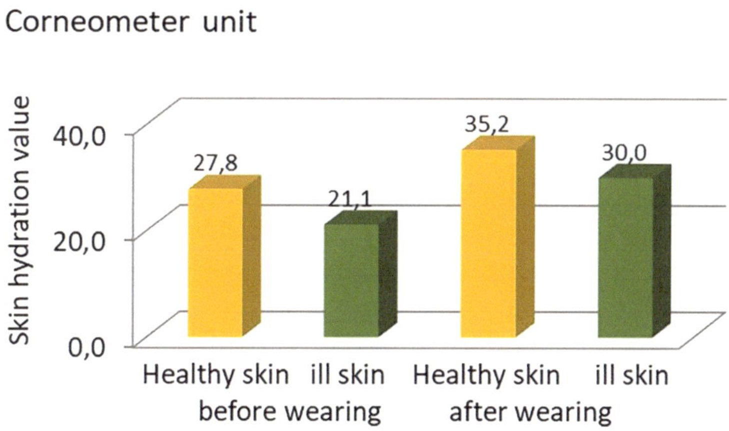

Figure 18.
Skin moisture content tested with the corneometric method (based on [45]).

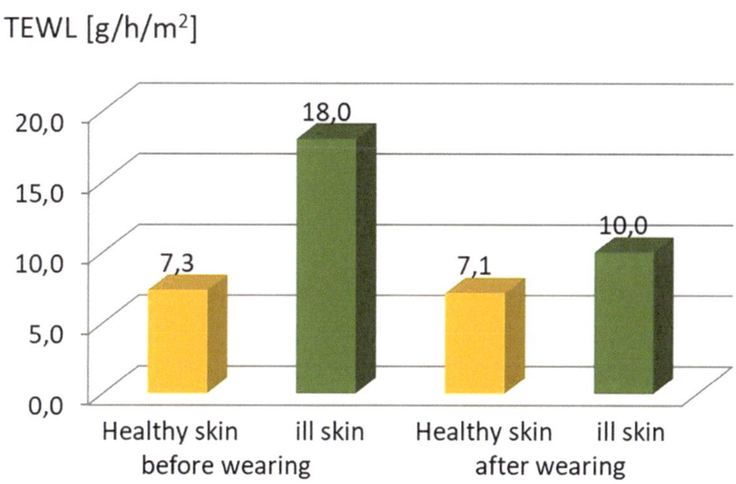

Figure 19.
Transepidermal water loss (TEWL) measured during experiment (based on [45]).

enriched with herbal extracts locked in microcapsules can serve as a supplement in the treatment of dermatological diseases.

7.2 Flax/hemp wound dressing

The study on the effective implementation of inherent antibacterial and anti-oxidant properties as well as high ability to liquid sorption of bast fibers for medical target resulted in the development of flax/hemp wound dressing. The investigation is protected by patent [46]. Carefully selected flax and hemp variety characterized by high phenolic acid content as well as the use of determined fiber processing that allows us to keep the bioactive substances in the fibers on proper level leads to obtain bioactive textile suitable to wound dressing when the structure of dressing is developed accordingly.

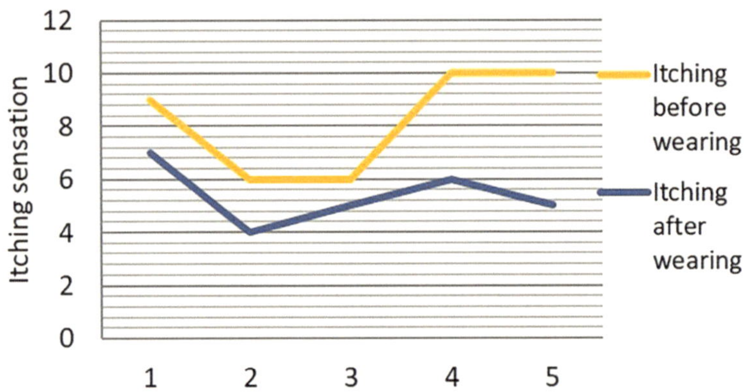

Figure 20.
Comparison of intensity of ill skin itching tested with a numeric scale in patients before and after wearing of the tested clothing (based on [45]).

The developed flax/hemp wound dressing was clinically tested with patients under care of surgical clinic suffering from nonhealing venous ulcers of legs and diabetic wounds. One of the examples of the efficiency of linen dressing is nonhealing venous ulcer of the right shin of the one patient. Before linen dress application, the patient was treated with using different medicines for three years with no success (antibiotics and ointment). After the application of treatment by the linen dress, the duration of healing was 4 months.

8. Conclusion

A multiperspective approach to bast fibers reveals their high potential to be used for the development and implementation of safe barrier and functional textiles. The chemical composition of the lignocellulosic fibers, including phenolic acids, results in their unique inherent bioactivity, which cannot be found in other natural or man-made fibers can give some competitive advantage of linen/hemp bioproducts. Owing to the sensitivity of phenolic acids on some wet aggressive methods of fiber extraction and processes, the design of technological chain should be well cherry-picked and focused on keeping the active fiber components on the desired level in order to produce bioactive functional textiles. Entering of other active substances on the textile surface, such as nanosilver, nanolignin, or simply herbal extracts, can strengthen the functionality of final linen/hemp goods. The linen/hemp apparels provide well-being and optimal comfort to the users, can cause improvement of skin condition, and protect against UV radiation. Bast fiber sector covering both agriculture and industry meets environmental requirements determined by the European Green Deal (EGD) strategy and contributes to the limitation of greenhouse gasses emission. Flax and hemp cultivation and their use for different purposes are in line with aims of EGD, particularly, can have influence on the improvement of the well-being and health of citizens as well as can support activity for future generations by caring for the environment and sustainability of the holistic value chain for bast fiber bioproducts.

Bast-fiber-based textile targets the improvement of human everyday life on the following levels:

- by direct covering body of wearer ensuring optimal comfort and healthy skin,
- by sustainability and lowering of environmental impact,
- by potential to improve bioeconomy,
- by possibility to use all by-products and be in line with the zero-waste strategy.

Acknowledgements

This work was realized within Project POIR.01.01.01-00-0597/21 entitled: Development of biodegradable face mask Type II from natural fibers, which is co-financed by European Regional Development Fund and The National Center for Research and Development in Poland.

Author details

Malgorzata Zimniewska* and Barbara Romanowska
Institute of Natural Fibres and Medicinal Plants National Research Institute, Poznan, Poland

*Address all correspondence to: malgorzata.zimniewska@iwnirz.pl

References

[1] European Commission [Internet]. 2019. Available from: https://ec.europa.eu/info/strategy/priorities-2019-2024/european-green-deal_en

[2] Turunen L, van der Werf HMG. The production chain of hemp and flax textile yarn and its environmental impacts. Journal of Industrial Hemp. 2007;**12**(2):43-66. DOI: 10.1300/J237v12n02_04

[3] Van der Werf HMG, Turunen L. The environmental impacts of the production of hemp and flax textile yarn. Industrial Crops and Products. 2008;**27**(1):1-10. DOI: 10.1016/j.indcrop.2007.05.003

[4] Le Duigou A, Davies P, Baley C. Environmental impact analysis of the production of flax fibres to be used as composite material reinforcement. Journal of Biobased Materials and Bioenergy. 2011;**5**(1):153-165. DOI: 10.1166/jbmb.2011.1116

[5] Deng Y, Tian Y. Assessing the environmental impact of flax fibre reinforced polymer composite from a consequential life cycle assessment perspective. Sustainability. 2015;**7**(9):11462-11483. DOI: 10.3390/su70911462

[6] Zimniewska M. Hemp fibre properties and processing target textile: Review. Materials. 2022;**15**(5):1901. DOI: 10.3390/ma15051901

[7] Labouze E, Le Guern Y, Petiot CH. Eco-profile of a linen shirt and comparative analysis of linen and cotton shirts. In: Labouze E, Le Guern Y, Petiot CH, editors. Organisations professionnelles du lin: Bio Intelligence Service S.A.S. 2007. pp. 1-9. http://news.europeanflax.com/wp-content/uploads/2020/05/ACV-_-VERSION-EN.pdf

[8] Härkäsalmi T. Innovaatiohakuisuus runkokuitujen tekstiilituotannossa – konseptoinnilla kohti lyhytkuitujen tuotteistamista. Faculty of Agriculture and Forestry, Doctoral dissertation. University of Helsinki. Dissertation. 2006

[9] Nykter M. Microbial Quality of Hemp (*Cannabis sativa L.*) and Flax (*Linum usitatissimum L.*) from plants to thermal insulation. Faculty of Agriculture and Forestry, Department of Agrotechnology. Doctoral dissertation. Helsinki: University of Helsinki; 2006. ISBN 952-10-3208-1

[10] Lefeuvre A, Bourmaud A, Lebrun L, Morvan C, Baley C. A study of the yearly reproducibility of flax fiber tensile properties. Industrial Crops and Products. 2013;**50**:400-407. DOI: 10.1016/j.indcrop.2013.07.035

[11] Beyette B. 100% Wild Organic Linen/Flax Fiber Directly From Belarus [Internet]. 2022. Available from: https://pl.pinterest.com/pin/53832158018247849/

[12] Szałkowski Z, editor. Poradnik inżyniera. Włókiennictwo. Warszawa: Wydawnictwo Naukowo-Techniczne; 1978. p. 1247

[13] Gibson LJ. The hierarchical structure and mechanics of plant materials. Journal of the Royal Society Interface. 2013;**9**(76):2749-2766. DOI: 10.1098/rsif.2012.0341

[14] Urbańczyk G. Nauka o Włóknie. Warszawa: WNT; 1985

[15] John MJ, Anandjiwala RD. Recent developments in chemical modification

and characterization of natural fiber-reinforced composites. Polymer Composites. 2008;**29**(2):187-207. DOI: 10.1002/pc.20461

[16] Franck RE. Bast and other plant fibers. Cambridge: Woodhead Publishing Limited; 2000. pp. 176-206. ISBN 1-85573-684-5

[17] Zimniewska M, Władyka-Przybylak M. Natural fibers for composite applications. In: Rana S, Fangueiro R, editors. Fibrous and Textile Materials for Composite Applications. Singapore: Textile Science and Clothing Technology; 2016. pp. 171-204. DOI: 10.1007/978-981-10-0234-2_5. ISBN 978-981-10-0234-2

[18] Bismarck A, Aranberri-Askargorta I, Springer J, Lampke T, Wielage B, Stamboulis A, et al. Surface characterization of flax, hemp and cellulose fibers; surface properties and the water uptake behavior. Polymer Composites. 2002;**23**(5):872-894. DOI: 10.1002/pc.10485

[19] Zimniewska M, Rożańska W, Gryszczyńska A, Romanowska B, Kicińska-Jakubowska A. Antioxidant potential of hemp and flax fibers depending on their chemical composition. Molecules. 2018;**23**(8):1993. DOI: 10.3390/molecules23081993

[20] Daenekindt A. Flax, hemp, and allied fibres in the world. Euroflax Newsletter. 2004;**21**(1):6-9

[21] Akin DE. Linen Most useful: Perspectives on structure, chemistry, and enzymes for retting flax. ISRN Biotechnology. 2012;**2013**:186534. DOI: 10.5402/2013/186534

[22] Jankauskienė Z, Butkutė B, Gruzdevienė E, Cesevičienė J, Fernando AL. Chemical composition and physical properties of dew- and water-retted hemp fibers. Industrial Crops and Products. 2015;**75**(B):206-211. DOI: 10.1016/j.indcrop.2015.06.044

[23] Zimniewska M, Zbrowski A, Konczewicz W, Majcher A, Przybylski J, Matecki K, et al. Cottonisation of decorticated flax fibres. FIBRES & TEXTILES in Eastern Europe. 2017;**25**(3(123)):26-33. DOI: 10.5604/01.3001.0010.1685

[24] Mustata A, Mustata FSC. Moisture absorption and desorption in flax and hemp fibres and yarns. Fibres & Textiles in Eastern Europe. 2013;**21**(3(99)):26-30

[25] Pouillot A, Polla LL, Tacchini P, Neequaye A, Polla A, Polla B. Chapter 13: Natural antioxidants and their effects on the skin. In: Dayan N, Kromidas L, editors. Formulating, Packaging, and Marketing of Natural Cosmetic Products. 1st ed. New York: John Wiley & Sons, Inc.; 2011. pp. 239-257. DOI: 10.1002/9781118056806.ch13

[26] Cotoras M, Vivanco H, Melo R, Aguirre M, Silva E, Mendoza L. In vitro and in vivo evaluation of the antioxidant and prooxidant activity of phenolic compounds obtained from grape (Vitis vinifera) pomace. Molecules. 2014;**19**(12):21154-21167. DOI: 10.3390/molecules191221154

[27] Sinapic CC. Acid and its derivatives as medicine in oxidative stress-induced diseases and aging. Oxidative Medicine and Cellular Longevity. 2016;**2016**:3571614. DOI: 10.1155/2016/3571614

[28] Gorshkova TA, Salnikov VV, Pogodina NM, Chemikosova SB, Yablokova EV, Ulanov AV, et al. Composition and distribution of cell wall phenolic compounds in flax (*Linum usitatissimum L.*) stem tissues. Annals

of Botany. 2000;**85**(4):477-486. DOI: 10.1006/anbo.1999.1091

[29] Iiyama K, Lam TBT, Stone BA. Covalent cross-links in the cell wall. Plant Physiology. 1994;**104**(2):315-320. DOI: 10.1104/pp.104.2.315

[30] Zimniewska M. Antioxidant activity of Fibres originating from traditional varieties of polish flax plants. Fibres & Textiles in Eastern Europe. 2015;**23**(6(114)):41-47. DOI: 10.5604/12303666.1167418

[31] Zimniewska M, Goślińska-Kuźniarek O. Evaluation of antibacterial activity of flax fibers against the staphylococcus aureus bacteria strain. FIBRES & TEXTILES in Eastern Europe. 2016;**24**(2):120-125. DOI: 10.5604/12303666.1191439

[32] Zuk M, Dorotkiewicz-Jach A, Drulis-Kawa Z, Arendt M, Kulma A, Szopa J. Bactericidal activities of GM flax seedcake extract on pathogenic bacteria clinical strains. BMC Biotechnology. 2014;**14**:70 http://www.biomedcentral.com/1472-6750/14/70

[33] Racu C, Cogeanu AM, Diaconescu RM, Grigoriu A. Antimicrobial treatments of hemp fibers grafted with β-cyclodextrin derivatives. Textile Research Journal. 2012;**82**(13):1317-1328. DOI: 10.1177/0040517512441995

[34] Emam HE, El-Rafie MH, Rehan M. Functionalization of unbleached flax fibers by direct integration of Nano-silver through internal and external reduction. Fibers and Polymers. 2021;**22**:3014-3024. DOI: 10.1007/s12221-021-0993-y

[35] Milanović J, Mihailović T, Popović K, Kostić M. Antimicrobial oxidized hemp fibers with incorporated silver particles. Journal of the Serbian Chemical Society. 2012;**77**(12):1759-1773. DOI: 10.2298/JSC121018143M

[36] Chang L, Duan W, Huang S, Chen A, Li J, Tang H, et al. Improved antibacterial activity of hemp fibre by covalent grafting of quaternary ammonium groups. Royal Society Open Science. 2021;**8**:201904. DOI: 10.1098/rsos.201904

[37] Zimniewska M, Krucińska I. The role of apparels based on natural fibres: Flax in human physiology and health. In: Kozłowski RM, Muzyczek M, editors. Natural Fibers: Properties, Mechanical Behavior, Functionalization and Applications. New York: Nova Science Publishers Inc; 2017. pp. 193-212. ISBN: 978-1-53612-071-4

[38] Zimniewska M, Krucińska I. The effect of raw material composition of clothes on selected physiological parameters of human organism. Journal of the Textile Institute. 2010;**101**(2):154-164. DOI: 10.1080/00405000802309568

[39] Zimniewska M, Huber J, Krucińska I, Torlińska T, Kozłowski R. The influence of clothes made from natural and synthetic fibers on the activity of the motor units in selected muscles in the forearm – Preliminary studies. FIBRES & TEXTILES in Eastern Europe. 2002;**10**(4):55-59

[40] Zimniewska M, Witmanowski H, Kozłowski R. Clothing effect on selected parameters of oxidative stress. Lenzinger Berichte. 2006;**85**:17-21

[41] Zimniewska M, Batog J. Chapter 4: Ultraviolet-blocking properties of natural fibres handbook of natural fibres. In: Kozłowski RM, editor. Volume 2: Processing and Applications. Cambridge: Woodhead Publishing Limited; 2012. pp. 141-167. ISBN 978-1-84569-698-6

[42] Zimniewska M, Kozłowski R, Batog J. Nanolignin modified linen fabric as multifunctional product. Molecular Crystals and Liquid Crystals. 2008;**484**:43-50. DOI: 10.1080/15421400801903395

[43] Zimniewska M, Kozłowski R, Batog J. Cellulose fibre textiles containing nanolignin, a method of applying nanolignin onto textiles and the use of nanolignin in textile production. European Patent No. EP 2150649 B1. 2012

[44] Zimniewska M, Krucińska I, Frydrych I, Mikołajczak P, Schmidt-Przewoźna K, Pawlaczyk M, et al. Clothing dressing as a supplement in the treatment of dermatological diseases. Patent No. 229203. 2018

[45] Zimniewska M, Pawlaczyk M, Krucińska I, Frydrych I, Mikołajczak P, Schmidt-Przewoźna K, et al. The influence of natural functional clothing on some biophysical parameters of the skin. Textile Research Journal. 2019;**89**(8):1381-1393. DOI: 10.1177/0040517518770680

[46] Zimniewska M, Szkaradkiewicz A, Szlęzak A, Spychalski G, Władyka-Przybylak M. Wound dressing made of natural bast fibers flax/hemp, method of their manufacture and application natural genetically unmodified bast fibers flax/ hemp to wound dressing manufacture. Patent No. PL 220546. 2015

Section 3

Characterization and Analysis of Natural Fibers

Chapter 4

Hydro/Hygrothermal Behavior of Plant Fibers and Its Influence on Bio-Composite Properties

Nouri Mustapha and Tahlaiti Mahfoud

Abstract

Plant fibers have been shown to be highly sensitive to water molecules; this impacts the functionality of composites reinforced with these fibers, commonly known as bio-composites. This review aims to provide a comprehensive description of the behavior of plant fibers in the presence of water molecules in a liquid or gaseous state, as well as the different phenomena and mechanisms involved at the fiber scale and at the bio-composite scale via recent studies in this field. First, we will discuss the physical problem of sorption in polymers in a general way, and then we will focus on the case of plant fibers. Particular attention will be given to the adsorption kinetics of plant fibers and the models used to determine their diffusion parameters. In a second step, the effect of the incorporation of plant fibers in polymer matrices will be examined as well as the different factors influencing the diffusive behavior of bio-composites. In addition, the effect of hydro/hygrothermal aging on the mechanical properties of bio-composites will be discussed.

Keywords: plant fibers, bio-composite, hygrothermal behavior, aging, diffusion

1. Introduction

The depletion of natural resources, the pollution caused by industry, and the current environmental challenge have led researchers and industrialists to explore new ways of providing green and renewable resources, such as biomass. Among these promising natural resources in the field of composite materials are plant fibers, also known as cellulosic or lignocellulosic fibers. Indeed, the integration of these fibers into the field of composite materials as an alternative to synthetic fibers continues to attract much research and application interest. This is due to their neutrality with regard to CO_2 emissions into the atmosphere, but also to the fact that these fibers have specific mechanical and thermal properties. In addition, these fibers have other advantages, such as their low cost, their biodegradability, and the fact that they are renewable.

However, the hydrophilic nature of these plant fibers and the diversity of their properties represent limitations for their use as reinforcement for composite materials. This is particularly true of hydrophobic matrices, which raises questions about the compatibility between the reinforcement and the matrix on the one hand, and the

IntechOpen

durability of the bio-composite on the other, especially when the latter is exposed to changing hygrothermal conditions.

This chapter will comprehensively address the hydro/hygrothermal behavior of plant fibers and their influence on the durability of bio-composites. First, we will discuss the physical problem of sorption in polymers in a general way, and then we will focus on the case of plant fibers. Particular attention will be given to the adsorption kinetics of plant fibers and the models used to determine their diffusion parameters. In a second step, the effect of the incorporation of plant fibers in polymer matrices will be examined as well as the different factors influencing the diffusive behavior of bio-composites. In addition, the effect of hydro/hygrothermal aging on the mechanical properties of bio-composites will be discussed.

Before starting this chapter, it is necessary to define some basic knowledge that are essential for understanding the physical phenomena that will be presented.

2. Generality on hydro/hygrothermal transfer

2.1 Terminology

2.1.1 Adsorption, absorption, and sorption

Absorption is a phenomenon of filling the pores of a solid with liquid water; it occurs by capillary condensation. Whereas adsorption is a process of attachment of water molecules, water vapor, to the surface of materials by intermolecular forces. These forces are of the same nature as those responsible for the imperfection of real gases and the condensation of vapors [1]. We, therefore, talk about the absorption of liquid water (hydrothermal aging) and the adsorption of water vapor (hygrothermal aging). As for sorption, it encompasses both phenomena.

2.1.2 Relationship between relative humidity and water activity

Under isothermal conditions, the water activity (a_w) in a material is defined by the ratio of the water vapor pressure at its surface (p_{vs}) to the pure water vapor pressure (p_{ve}). When the material is in equilibrium with its environment or there is no mass or heat transfer, p_{vs} equals the water vapor pressure of air (p_{va}). Therefore, the water activity of a material in equilibrium with its environment is equal to the relative humidity of the air (RH_a).

$$a_w = HR_a = \frac{P_{vs}}{P_{ve}(T)} = \frac{P_{va}}{P_{ve}(T)} \tag{1}$$

2.1.3 Condensation capillaire

This is the phenomenon where water vapor condenses into a liquid phase in a pore at an a_w of less than 1, where a_w = 1 represents the case of pure water.

2.1.4 Sorption isotherms

Under isothermal conditions, the curve representing the equilibrium moisture content (EMC) of a material as a function of a_w is called the sorption isotherm if it was

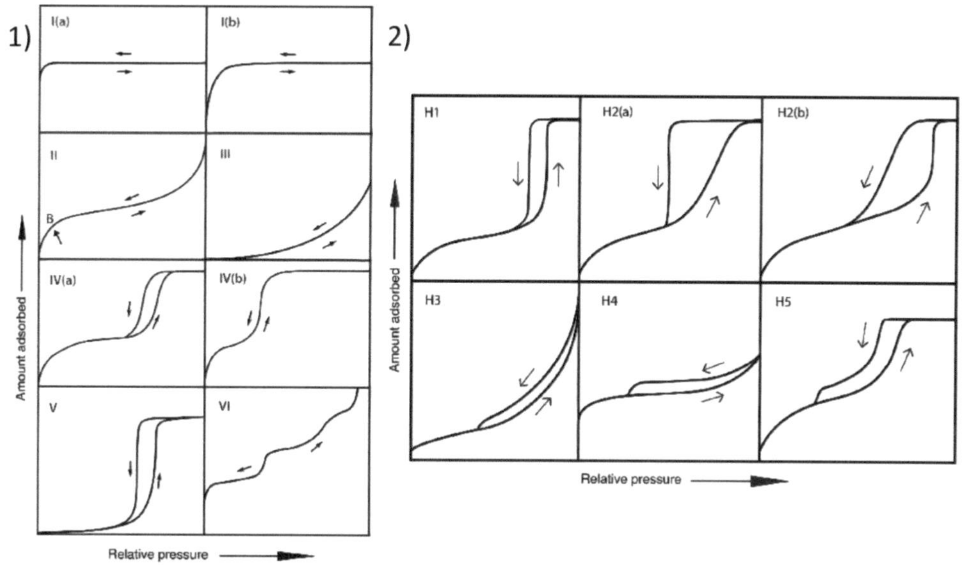

Figure 1.
IUPAC classification of physisorption isotherms (1) and hysteresis loops (2) [1].

determined from the dry state of the material, and the desorption isotherm if the material was saturated at the initial state [2].

Isotherms are classified according to the new IUPAC classification [1] into six types (**Figure 1**(1)). The shape of the isotherm provides information about the pore size (macropores, mesopores, or micropores) and the types of adsorption (monolayer or multilayer). When the desorption process is different from the sorption process, a hysteresis loop is observed. These loops are also classified by IUPAC [1] into five categories (**Figure 1**(2)).

2.1.5 Pore size

According to the IUPAC pore classification [1], macropores, mesopores, and micropores are, respectively, those pores whose width exceeds 50 nm, between 2 and 50 nm, and do not exceed 2 nm, respectively.

2.1.6 Aging

Aging is defined according to Berges [3] as an evolution of one or more properties of the material through a modification of the structure, the composition, or the morphology of the constituents. These changes can be temporary and dependent on the presence of the aging source (reversible aging), or permanent (irreversible aging) [4]. Furthermore, the ability of a material to resist irreversible aging is defined by its durability.

2.2 Sorption and diffusion of water molecules in polymers

Generally, polymers are penetrable by water molecules. The latter propagate progressively through the polymer macromolecular network when it is in direct contact with water molecules (gaseous or liquid) [5]. Over time, this leads to a weight gain that continues until the material reaches a saturation plateau [5]. This adsorbed amount is related to the total amount of hydrophilic sites, polymeric areas sensitive to receive water molecules, available in the polymer chains [6]. This also means that the chemical potentials of water in the polymer and in the ambient environment are equal [7]. The kinetics of water sorption and the amount of water adsorbed/absorbed depends on the polymer nature (hydrophobic or hydrophilic), the characteristics of the water (pH, deionized or salted water), and other thermodynamic parameters [4].

The most common method for evaluating water sorption processes in polymers is the recording of mass gain versus time data. The gravimetric curve represents the plot of this data; the mass of water absorbed/adsorbed (M_t) versus time (LF curve in **Figure 2**). This curve contains important information: the linear part of this curve informs about the penetration rate of water (the diffusion), while the saturation level presents the mass adsorbed/absorbed at infinity (M_∞) [7].

The diffusion behavior of polymers is often based on the Fick model. Nevertheless, some materials may exhibit non-Fick behavior in the presence of anomalies (curves A B, C, and D in **Figure 2**). The diffusion behavior can be identified mathematically using Eq. (2) [9]. If the parameter n is close to 0.5, this indicates that the diffusion of water, in this case, follows Fick's law.

$$\frac{M_t}{M_\infty} = kt^n \tag{2}$$

Where k, n represents the diffusion kinetic parameters.

In the case of bio-composites, we can see in the literature that the gravimetric curve of this kind of material shows a Fickian behavior at the macroscopic scale, see Section 4. However, at the microscopic scale, the problem is more complicated, water molecules can penetrate bio-composites by three different mechanisms: between polymer chains (matrix and fiber), by capillary action in micro-voids, and at the interface level (between fibers and the matrix) when chemical adhesion is absent.

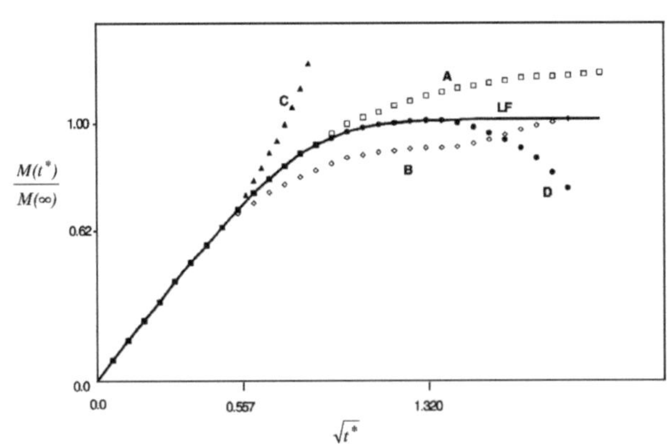

Figure 2.
Typical gravimetric absorption curves of a fluid: (LF) Fick behavior, (A) pseudo-Fick behavior, (B) 2-step diffusion, (C) diffusion with mechanical damage, and (D) diffusion with chemical damage [8].

2.3 Fick's mechanism

Fick's laws were established by the analogy between conductive heat transfer and mass transfer [10]. In an isotropic case, Fick's second law is given by the equation below:

$$\frac{\partial C}{\partial t} = D\left(\frac{\partial^2 C}{\partial x^2} + \frac{\partial^2 C}{\partial y^2} + \frac{\partial^2 C}{\partial z^2}\right) \tag{3}$$

Where: D is the diffusivity coefficient mm^2/s, it is a scalar that defines the diffusion kinetics. C is the equilibrium water concentration.

The above equation could also be written in a case of radial diffusion in a solid cylinder of radius r as follows:

$$\frac{\partial C}{\partial t} = D\left(\frac{1}{r}\frac{\partial C}{\partial r} + \frac{\partial^2 C}{\partial r^2}\right) \tag{4}$$

In the case where a uniform concentration is imposed on the borders, the simplified solution for long times ($0 < M_t/M_\infty < 0.5$) and short times ($0.5 < M_t/M_\infty < 1$) of the above relationship becomes, respectively, as follows [11]:

$$\frac{M_t}{M_\infty} = \frac{4}{\sqrt{\pi}}\tau^{1/2}\left[1 - \frac{\sqrt{\pi}}{4}\tau^{\frac{1}{2}} - \frac{\tau}{12}\dots.\right] \tag{5}$$

$$\frac{M_t}{M_\infty} = 1 - \sum_{n=1}^{\infty}\frac{4}{\propto_n^2}\,exp\left(-\alpha_n^2\tau\right) \tag{6}$$

With $\tau = Dt/r^2$ and α_n the solutions of the Bessel equation of order n.

Equations (5) and (6) can be decomposed even more precisely according to the M_t/M_∞ intervals as follows [12]:

$0 < M_t/M_\infty < 0.2$:

$$\frac{M_t}{M_\infty} = \frac{4}{\sqrt{\pi}}\tau^{1/2} \tag{7}$$

$0.2 < M_t/M_\infty < 0.5$:

$$\frac{M_t}{M_\infty} = \frac{4}{\sqrt{\pi}}\tau^{1/2}\left[1 - \frac{\sqrt{\pi}}{4}\tau^{\frac{1}{2}} - \frac{\tau}{12}\dots.\right] \tag{8}$$

$0.5 < M_t/M_\infty < 0.7$:

$$\frac{M_t}{M_\infty} = 1 - 4\left(\frac{expexp\left(-\propto_1^2\tau\right)}{\propto_1^2} + \frac{expexp\left(-\propto_2^2\tau\right)}{\propto_2^2}\right) \tag{9}$$

$0.7 < M_t/M_\infty < 1$:

$$\frac{M_t}{M_\infty} = 1 - 4\left(\frac{expexp\left(-\propto_1^2\tau\right)}{\propto_1^2}\right) \tag{10}$$

With $\alpha_1 = 2.40483et\alpha_n = 5.52008$.

For the case of a plane plate of thickness h with a uniform initial concentration on both surfaces, the approximate analytical solution of Fick's law (Eq. (3)) for unidirectional diffusion along the thickness is expressed as follows [12]:

For the first half-absorption ($M_t/M_\infty < 0.5$):

$$\frac{M_t}{M_\infty} = \frac{4}{h}\sqrt{\frac{Dt}{\pi}} \tag{11}$$

For the second half-absorption ($M_t/M_\infty \geq 0.5$):

$$ln\left(1 - \frac{M_t}{M_\infty}\right) = ln\,\frac{8}{\pi^2} - \frac{\pi^2 Dt}{h^2} \tag{12}$$

This is often used to illustrate the transfer of water vapor in continuous media and has been adopted by several authors to simulate the transfer of liquid water in materials assumed to be continuous at the macroscopic scale such as bio-composites and plant fibers, see Sections 3 and 4.

3. Hydro/hygrothermal aging of natural fibers

Plant fibers have been shown to be highly sensitive to water molecules, which impacts the functionality of bio-composites [13]. This phenomenon is often related to the morphology of these fibers, their cavity (lumen), and the free hydroxyl groups present on their surface (hydrophilic sites) [13–16]. The components of plant fiber responsible for its hydrophilic character are mainly hemicellulose and pectin [7], although lignin and amorphous cellulose are also hydrophilic: lignin contains fewer -OH groups than the polysaccharides [17], while cellulose is less accessible. Therefore, a change in hygrothermal conditions could affect the degree of crystallinity of natural fibers, their stiffness, dimensional stability (swelling), and tensile strength [13, 17].

Table 1 shows the results of the saturation rates of the plant fibers after immersion in water. These results depend mainly on the nature of the plant species, the extraction method used, the application of the treatments, and the test protocol applied.

Furthermore, it appears that the hydrophilic character of natural fibers could be reduced by applying treatments, such as chemical and physical treatments [18–20, 23–48]. The reasons for these improvements are multiple and depend on the nature of the treatment used and its effect on the plant fibers. Chemical treatments aim to change the chemical composition of the fiber. For example, NaOH treatment removes some of the hydrophilic non-cellulosic materials such as lignin, hemicellulose, and pectin that cover the natural fiber interface. Also, free OH groups can react with the NaOH molecule forming fiber-cell-O-Na groups, which reduces the hydrophilic hydroxyl groups. The silane molecule, on the other hand, has an end containing alkoxy groups that can react with the hydroxyl groups of the plant fiber, giving it hydrophilic properties on its surface. The same mode of functioning for treatment with acetic acid, acrylic acid, and benzoyl [24, 46, 47]. Physical treatments such as Corona [31–33], Plasma [34–36], and thermal [18, 37, 38] mainly modify the structural and interfacial properties of the plant fibers without significantly changing their chemical composition.

Fibers	Saturation rate (%)	References
Flax	42.58	Stamboulis et al. [18]
Alfa	140.60	Bessadok et al. [19]
Agave	138	Bessadok et al. [20]
Hemp	62	Saikia [21]
Hemp	63	Célino et al. [22]
Jute	67.8	
Lin	62.5	
Sisal	60.6	
Diss	40	Nouri et al. [23]

Table 1.
Saturation rates of the plant fibers after immersion in water.

The hydrophilic character of plant fibers strongly influences the hydro/hygrothermal properties (saturation rates, diffusion kinetics, and dimensional deformation) of bio-composites at the macroscopic scale, an increase in the fiber incorporation rate often leads to a change in these properties, see Section 4.

3.1 Sorption isotherm for plant fibers

The isotherm of cellulosic fibers generally has a sigmoidal shape with hysteresis loops between the adsorption and desorption curve [16, 17, 20, 25, 49–52], in accordance with the type II isotherm according to the "International Union of Pure and Applied Chemistry (IUPAC)" classification [1]. Since plant fibers are much more hydrophilic than polymer matrices, bio-composites generally exhibit the same type of isothermal sorption curve.

The origin of the hysteresis phenomenon is not yet fully understood and is still a subject of debate in the literature. Considering fiber cell walls as micro-mesoporous materials, some authors have linked this to capillary condensation, which occurs at high RH, and which could also be present at low RH in the micropores [51]. However, it has been shown that capillary condensation could only occur for such material at very high RH [53]. Others have suggested that this may be related to the change of state of the amorphous components, notably hemicellulose and lignin [50, 52, 54]. It has been found that when adsorption occurs above the glass transition temperature of these polymers, also known as the softening point, hysteresis should be absent or minimized [50, 52]. Indeed, at room temperature, amorphous cell wall components of cellulosic fibers such as wood could soften around 65–75% RH [50]. Keating et al. [54] reported a loss of sorption hysteresis in artificial hemicellulose (galactomannan) films when the RH is above 75% at 25°C (**Figure 3**). Furthermore, Salmèn and Larsson [50] found that increasing the temperature and decreasing the crystallinity index of the fibers affects the presence of the hysteresis loop (**Figure 4**). They also noticed that the swelling effect could be involved when RH is below the softening temperature. A priori, when water molecules penetrate the matrix, nanopores could be created in the structure to receive these molecules and under desorption conditions, these nanopores could collapse [54]. In addition, Hill et al. [55] questioned the lignin content, whose magnitude of sorption hysteresis was greater when the content of this component was high.

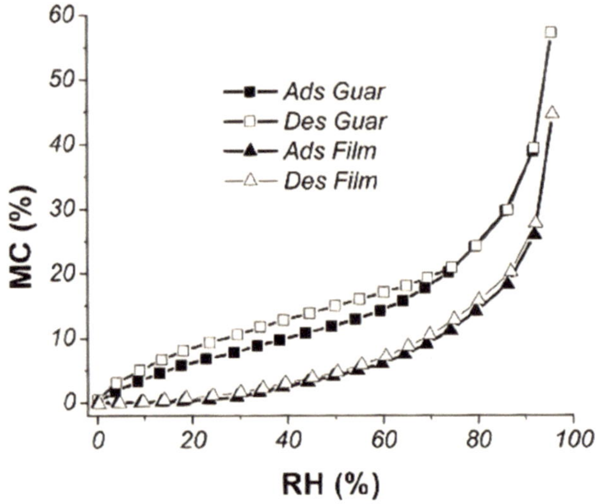

Figure 3.
Sorption isotherm of a nanocomposite film and a guar film as a function of RH, showing hysteresis [54].

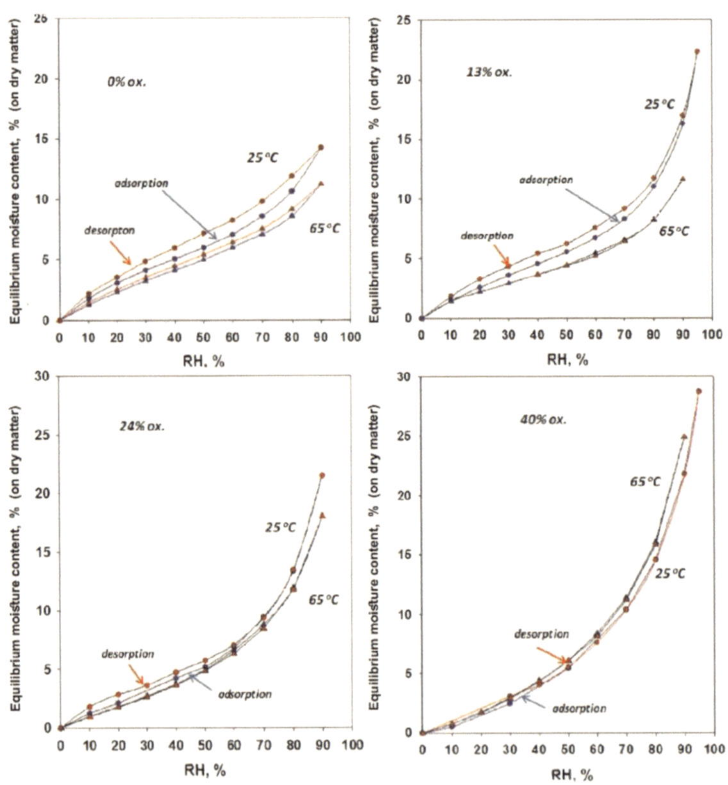

Figure 4.
Moisture adsorption and desorption curves as a function of RH for cellulose samples with different degrees of modification. Measurements were carried out at 25 and 65°C. The higher the degree of oxidation (ox), the lower the crystallinity [50].

In addition, several mathematical models exist in the literature to describe the isotherm of this kind of material, including theoretical, semi-theoretical, and empirical models [56]. The Guggenheim, Anderson, and de Boer (GAB), Hailwood Horrobin (H-H), and Generalized D'Arcy and Watt (GDW) models are the most widely used in the literature for plant fibers [20, 25, 55, 57–62] (**Table 2**). These models also make physical sense for the attachment of water molecules at the pore scale and can be explained by an extension of Langmuir's theory for multilayer adsorption. Indeed, each adsorption site can only adsorb one water molecule. These adsorbed molecules could subsequently be secondary adsorption sites for subsequent molecules. The GDW model assumes that some of these sites have the potential to become secondary adsorption sites. The H-H model, on the other hand, considers that water adsorbed by the cell wall can exist in two forms: multilayer water (M_d) and monolayer water (M_h). However, while the monolayer in the GAB and DGW model is invariant, the monolayer described in the H-H model can change over the entire RH range.

3.2 Water adsorption kinetics of plant fibers

The diffusive behavior of plant fibers are often based on the Fick model [10, 12, 18, 21, 22, 25, 57, 63]. However, it is sometimes described by non-Fick diffusion models. Célino et al. [22] used different classical diffusion models to define this phenomenon for plant fibers: the Fick model, the two-stage Fick model developed by Loh et al. [64], and the Langmuir model developed by Kibler et al. [65]. They concluded that the Fick model better represents the kinetics of water vapor adsorption in the studied fibers. However, the Langmuir model seems to fit better in the case of liquid water absorption during the immersion process (**Figure** 5(1)). In addition, Saikia [21] observed during his work on hemp, okra, and betel nut fibers that absorption behaved in two stages. The first stage took place very quickly and obeyed Fick's law of diffusion. The second absorption step represents a non-Fickian diffusion (**Figure** 5(2)).

On the other hand, the analytical solution of Fick's law in the case of plant fibers are approximate, the sorption is generally subdivided into two or even four zones, each of which is defined by its own law of behavior (see Eqs. (7)–(10)). Therefore, some authors propose a single diffusion coefficient to describe the diffusive behavior within plant fibers [18, 63], but others suggest two different diffusion coefficients

Modèles	Equation	Paramètres
Guggenheim, Anderson and de Boer (GAB).	$EMC_{GAP}(a_w) = \frac{km_0ca_w}{(1-ka_w)(1-ka_w+kca_w)}$	m_0 the adsorption capacity in the first layer in contact with the solid, c, and k are parameters related to the sorption enthalpy.
Generalized D'Arcy and Watt (GDW).	$EMC_{GDW}(a_w) = \frac{km_0a_w}{1+ka_w}\frac{1-k(1-\omega)a_w}{1-ka_w}$	ω represents the molecules adsorbed rate on primary sites which transform into secondary adsorption sites.
Hailwood and Horrobin (HH) model.	$EMC_{H-H}(a_w) = \frac{18}{W}\left(\frac{K_1K_2a_w}{1+K_1K_2a_w}\right) + \frac{18}{W}\left(\frac{K_2a_w}{1+K_2a_w}\right)$	W is the molecular the dry fiber weight per mole of sorption sites, K_1 and K_2 are constants.

Table 2.
Mathematical description of the GAB, GDW, and HH models.

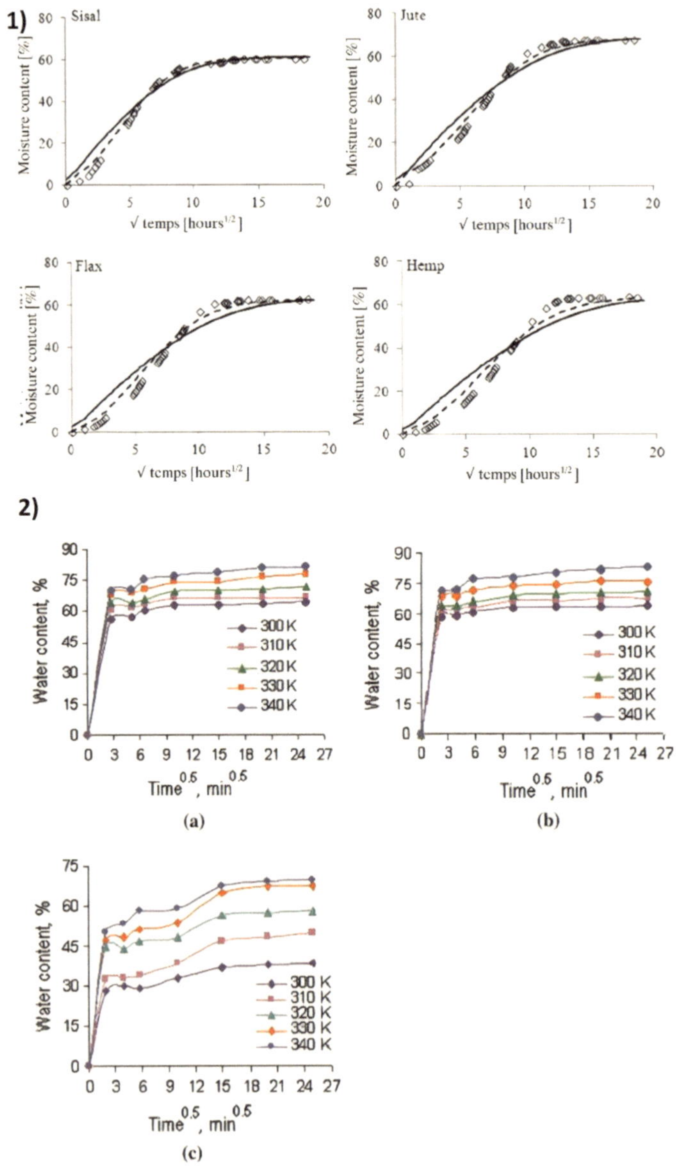

Figure 5.
(1) Immersion absorption curve for different plant fibers [22]. (2) Water content absorbed by: (a) hemp, (b) okra, and (c) betel nut as a function of the square root of time at different temperatures [21].

[12, 25, 57]: D_1 for the first half-sorption and D_2 for the second half-sorption. Gouanvé et al. [12] found during their work that the D_1 and D_2 values of flax fibers are similar throughout the a_w range studied. However, Alix et al. [25] observed a distinct behavior on the same type of fiber, the value of D_2 was found to be significantly larger than D_1 over the whole range of a_w studied. According to the authors, this dissimilarity is caused by the heterogeneity of the fibers where D_2 should be more representative of water diffusion in the core of the fiber while D_1 characterizes diffusion through the surface. The same findings were raised by Bessadok et al. [57] on Agave fibers and Nouri et al. [23] on Diss fibers.

In addition, a multitude of results on the diffusion coefficients of plant fibers have been found in the literature (**Table 3**), this is due mainly to the adopted test protocol, especially for the case of immersion in water, and to the assumed hypothesis (the diffusion occurs from the cross-section and the latter is assumed to be circular). Célino et al. [22] studied the sorption mechanism of different plant fibers (flax, hemp, sisal, and jute) under different conditions: immersion in water and conditioned at 80% RH. The results showed, for each type of fiber, a significant difference of the order of 10^{-2} between the diffusion coefficients of each case studied. For example, flax fibers showed a diffusion coefficient in the case of water immersion of 5.9 e^{-06} mm^2/s and 2.00 e^{-04} mm^2/s under 80% RH. In the case of relative humidity conditions, similar results were also obtained by Rouidier [10] on flax fibers. However, Gouavné et al. [12] noted a difference of more than three decades under the same conditions. Furthermore, in the case of water immersion, Stamboulis et al. [18] noted a D value of the same order of magnitude as those found by Célino et al. [22] and Rouidier [10] under relative humidity for flax fibers. On the other hand, Nouri et al. [23] studied the effect of different treatments on the diffusive behavior of Diss fibers under different conditions. The results showed diffusion coefficients of the

Fibers	Condition	Diffusion model	D (mm^2/s)	References
Jute	51% RH	Fick	3.38 e^{-07}	Mannan and Talukder [63]
Flax	Water immersion	Fick	1.93 e^{-04}	Stamboulis et al. [18]
Flax	Under a wide range of RH	Fick	6.50 e^{-07}	Gouavné et al. [12]
Agave	57%	Fick	D_1 = 7.27 e^{-03} D_2 = 3.29 e^{-02}	Bessadok et al. [57]
	75%	Fick	D_1 = 4.55 e^{-03} D_2 = 2.24 e^{-03}	
	87%	Fick	D_1 = 7.64 e^{-03} D_2 = 2.51 e^{-03}	
Hemp	Water immersion at 300 K	Fick	5.20 e^{-04}	Saikia [21]
Flax	33% RH	Fick	2.06 e^{-04}	Roudier [10]
	50% RH	Fick	2.79 e^{-04}	
	75% RH	Fick	3.69 e^{-04}	
Hemp	Water immersion	Langmuir	5.6 e^{-06}	Célino et al. [22]
Jute		Langmuir	5.9 e^{-06}	
Flax		Langmuir	6.8 e^{-06}	
Sisal		Langmuir	9.1 e^{-06}	
Hemp	80% RH	Fick	2.27 e^{-04}	
Jute		Fick	4.02 e^{-04}	
Flax		Fick	2.00 e^{-04}	
Sisal		Fick	1.17 e^{-04}	
Diss	Water immersion	Fick	D_1 = 7.90 e^{-08} D_2 = 5.6 e^{-07}	Nouri et al. [23]

Table 3.
Diffusion parameters of plant fibers under different aging conditions.

same order of magnitude for the different case studies. As an example, fibers treated with 5% NaOH showed a D_1 when immersed in water and conditioned under a $RH = 68\%$ of 4.30×10^{-8} mm²/s and 1.6×10^{-8} mm²/s, respectively.

In addition, most analytical approaches consider the cross-sectional shape of vegetal fiber bundles as circular. However, this hypothesis is not necessarily applicable to all vegetal fibers as shown by the observations made on the different vegetal fibers [66–69]. Assuming that the diffusion occurs from the cross-section, the morphology of the cross-section can influence the results obtained when calculating the diffusion coefficient. Recently, Nouri et al. [70] have proposed a coupled (experimental-numerical) approach to improve the modeling of water diffusion through Diss considering two fibers geometries, one often suggested in the literature (circular) and the other one revealed by microscopy (ellipsoidal). The modeling was applied for untreated and treated Diss fibers to determine the diffusion coefficient. Moreover, a single diffusion coefficient was enough to describe diffusion behavior in the case of Diss fibers, contrary to what is practiced with analytical approaches. It was demonstrated that for the same fiber cross-section area, faster diffusion will occur on high perimeter fibers. This means that when the fibers are represented by an ellipsoidal section instead of a circular one for the same cross-sectional area, the diffusion coefficient is less important. Therefore, a significant decrease of 1.5–2.4 in the diffusion coefficient was observed for the fibers studied.

Concerning the evolution of diffusion coefficients as a function of the evolution of relative humidity, Gouanvé et al. [12] observed an increase in D_1 and D_2 of flax fibers when a_w is lower than 0.50; above this value, a decrease was observed for these coefficients (**Figure 6a**). These findings were explained, according to the authors, by the dual-mode of sorption, Langmuir, and Henry, in the first part of the isotherm as

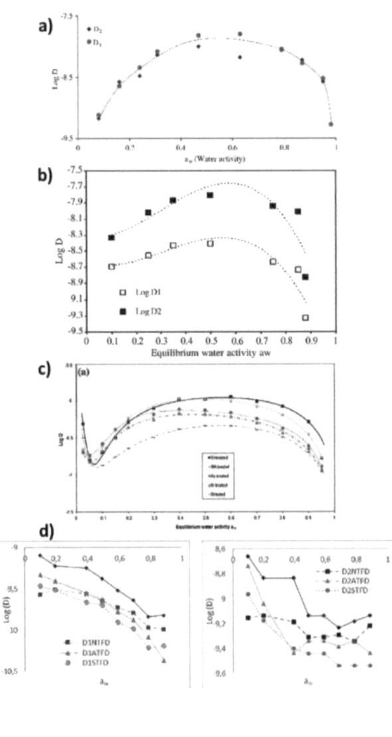

Figure 6.
Evolution of the water diffusion coefficients (D_1 *and* D_2) *as a function of* a_w *for: (a) flax fibers [12], (b) agave fibers [57], (c) flax fibers [25], and (d) Diss fibers [23].*

well as the braking effect of swelling at high a_w. The same findings were raised by Bessadok et al. [57] (**Figure 6b**). According to them, this indicates a water sorption mechanism according to Park's model: part of the water is adsorbed on specific sites (low mobility of the fixed water molecules) and the rest is dissolved according to Henry's process (higher mobility of the dissolved molecules) and then the subsequent formation of aggregates at high a_w (low mobility of the aggregates). Alix et al. [25] observed similar behavior for flax fibers with a decrease in diffusion coefficients when a_w is less than 0.1. Above this value, D_1, D_2 follow the behavior observed by Gouanvé et al. [12] (**Figure 6c**). This early decrease was explained by the fact that water molecules interact with polar fiber groups leading to hydrogen bonds that increase the cohesion between cellulose chains, thus reducing the mobility of water. In contrast to the previous authors, Nouri et al. [23] found a decrease in D (D_1, D_2) with increasing a_w (**Figure 6d**). On the other hand, Roudier [10] found that as the relative humidity increases, the diffusion coefficient for flax fibers increase. According to him, this relationship is described by a linear law (**Figure 7**). It should be noted that the fibers studied here were dried before each conditioning, contrary to the other works which are based on the results of isotherm adsorption.

Furthermore, treatments also have an important impact on the hygrothermal properties of plant fibers. Mannan et al. [63] found that after various treatments on jute fibers, the diffusion coefficients decreased during delignification, bleaching, and soap washing, suggesting that moisture was absorbed in the amorphous region of the fibers. Stamboulis et al. [18] studied the effect of the heat treatment, Durbalin, on the hygrothermal properties of flax fibers. The raw fibers always showed a higher diffusion coefficient than the treated fibers, independent of the *RH* studied. Bessadok et al. [57] observed a decrease in the diffusion coefficient of Agave fibers after their chemical treatment (acrylic acid, styrene, maleic anhydride, and acetylation). On the same line, Nouri et al. [23] noticed a decrease in the diffusion coefficient of Diss fibers for the water absorption case after the different treatments carried out (thermal, acetic acid, NaOH, and silane treatments).

The relationship between the diffusion properties of plant fibers and bio-composites is little studied in the literature due to the lack of reliable experimental protocols for plant fibers (for the case of immersion in water) on the one hand, and the complexity of the problem on the other hand. In the next section, we give more explanation about the effect of integrating plant fibers as reinforcement to the polymer matrix on the diffusive behavior of bio-composites.

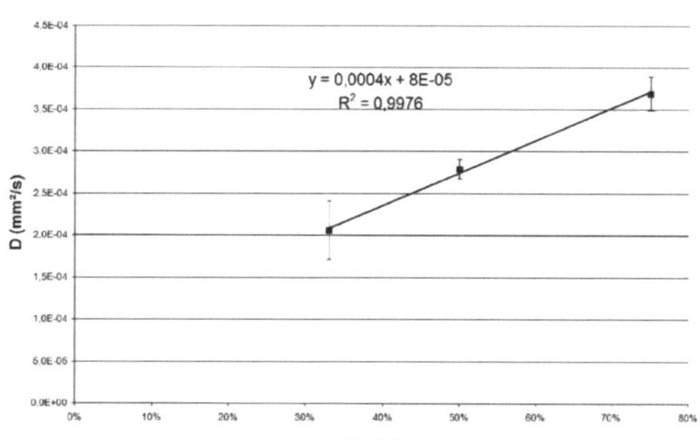

Figure 7.
Linear evolution of the water diffusion coefficient for flax fibers as a function of relative humidity [10].

4. Hydro/hygrothermal aging of bio-composites

In contrast to plant fibers, water sensitivity is not as noticeable in most polymer matrices, especially petroleum-based matrices. In this section, we investigate the effect of incorporating plant fibers on the hydro/hygrothermal behavior of bio-composites.

4.1 Diffusion behavior of bio-composites

We can see in the literature that most of the plant fiber/PP bio-composites show a Fickian diffusion: wood powder/polypropylene (PP) [71], jute fiber/PP [72, 73], date palm fiber/PP [74], hemp fiber/P and bagasse fiber/PP [75], Kenaf fiber/PP [76], wood fiber/PP [77], wood fiber/high density polyethylene (HDPE) [78, 79], flax fiber/Elium and flax fiber/epoxy [7], flax/epoxy fiber [3], short jute fiber/PLA [80], Areca fine fibers, and *Calotropis gigantea* fiber/phenol formaldehyde [81].

For the case of a flat plate of thickness h with infinite dimensions, the approximate analytical solution of Fick's law for unidirectional diffusion through the thickness is expressed by Eqs. (11) and (12). This is of course to determine the homogenized properties of the bio-composite at the macroscopic scale (effective properties). Furthermore, in the case of finite dimensions, the direction of diffusion can be perturbed by so-called edge effects; the diffusion is not only unidirectional along with the thickness but also occurs in other directions. Chilali [7] studied the influence of the width (w)/thickness (h) ratio on the water diffusion coefficient of Lin fiber/epoxy and Lin fiber/Elium composites. The results showed that when this ratio is equal to 60 ($180 \times 180 \times 3\ mm^3$) the edge effect is no longer observable and diffusion occurs along with the thickness similar to the case of the samples ($20 \times 20 \times 3\ mm^3$) sealed on the edges. Therefore, and in addition to what was explained in Section 2.3 for the plane plate case, Shen and Springer [82] proposed a correction to the diffusion coefficient (D_c) in the case of finite sample dimensions for a homogeneous material:

$$D_c = \frac{D}{\left[1 + \frac{h}{l} + \frac{h}{w}\right]^2} \tag{13}$$

This correction is often applied in the literature on bio-composites to avoid the edge effect [72, 83–85].

On the other hand, the interpretation of the diffusion process of composites by analytical approaches often does not reveal all the secrets of this phenomenon, especially when such a complex material is studied.

For a better understanding, the use of numerical approaches using the finite element method is frequently reported in the literature. We can distinguish two types of modeling of the diffusive behavior of conventional composites in the literature: those that consider the composite as a homogeneous material defined by its effective properties [86–88], others that represent it in a more realistic way by taking into account the fibers [80, 89–92]. These models can be made in 3D or 2D. In addition, the use of Fick's law is often observed [80, 86–91], although we can sometimes see the use of non-Fick's laws such as Langmuir's [89, 92, 93]. However, there is little work on bio-composites. Chilali [7] modeled the diffusive behavior of bidirectional flax fiber/epoxy and flax fiber/Elium bio-composites by a 2D biphasic model using the 2D Fick model. Berges [3] reproduced the same methodology for unidirectional flax/epoxy fiber composites. Jiang et al. [80] modeled the 3D diffusive behavior of short jute/PLA composites based on microstructure identification by X-ray tomography. Nouri [94] proposed a 2D numerical model for Diss/PP bio-composites considering three phases: fibers, matrix, and interface. This was done based on microscopic pictures of the bio-composite microstructure. This was achieved based on microscopic pictures of the bio-composite microstructure. The difficulty in this kind of modeling is the non-adaptation of the values determined at the fiber scale, notably the diffusion coefficient as explained in Section 3, to the value that should be implemented in the model. Generally, the authors need to recalculate the fiber diffusion coefficient by an inverse method.

4.2 Factors influencing the hydro/hygrothermal behavior of bio-composites

In general, water molecules penetrate bio-composites by three different mechanisms: between polymer chains, by capillary action in micro-voids, and in the interfaces between fibers and the matrix [81]. Therefore, the process of water diffusion through bio-composites is influenced, mainly, by two types of factors, namely internal factors (related to the bio-composite structure and the nature of its phases) and external factors (relative humidity and temperature).

The literature has shown that the fiber content influences the hydro/hygrothermal properties of bio-composites, Aloa et al. [28] studied the water absorption behavior of hemp fiber/PLA bio-composites and found that water absorption and swelling increased with the addition of hemp fibers to the PLA matrix. The same results were observed by Reza and Krishna [95], Sanjeevi et al. [81], and Mishra and Verma [96] for water absorption. The latter authors explained this, in their study of wood flour/PP composites, by the increase in free OH groups with increasing wood flour; this makes the bio-composite more hydrophilic. However, the diffusion coefficient does not seem to be impacted by increasing the loading rate, and the bio-composites show a lower diffusion coefficient than PP (a slowing wood flour effect). In contrast, results obtained by Law and Ishak [33] show that the diffusion coefficient of Kenaf/PP fiber composites increases with increasing loading rate (an accelerating Kenaf fiber effect). The authors explained this by the ability of the matrix to surround most of the fibers at low loading rates acting as a barrier to water diffusion. This barrier effect of the matrix decreases with increasing loading rate, leading to accelerated diffusivity. Similar results were reported by Joseph et al. [97] on Sisal/PP fiber composites.

On the other hand, a better fiber/matrix adhesion can lead to a reduction in the number of hydrophilic sites in the bio-composite and consequently influence its diffusive behavior. Beg and Pickering [98] found a decrease in the diffusion coefficient of kraft fiber/PP composites with a loading rate of 40% after the addition of 4% by weight MA-g-PP (coupling agent), see **Figure 8**. This decrease has also been observed by many researchers [76, 99]. However, the work of Mishra and Verma [96] has shown the opposite.

On the other hand, Sanjevvi et al. [81] found that fiber size had a significant influence on the adsorption rate of Areca fine fibers (AFFs) and *C. gigantea* fiber/phenol formaldehyde bio-composites whose long-fiber composite showed the highest water uptake of about 20% when the volume loading rate was 25%. In addition, Pérez-Fonseca et al. [100] found that the size of the fibers (short: 150–212 mm and long: 300–425 mm) did not have a significant effect on the water absorption kinetics of Agave/PP and Pine Sawdust/PP composites. Nevertheless, the composites containing Agave fibers had about three times higher absorption than the composites loaded with

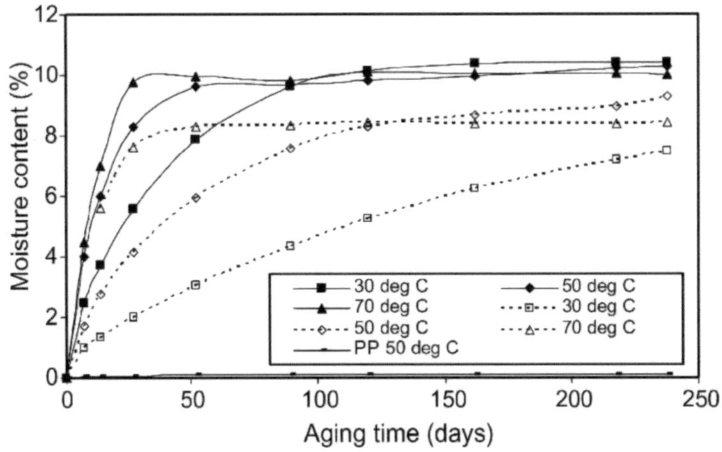

Figure 8.
Evolution of moisture content of composites (kraft fiber/PP) as a function of water immersion time during hygrothermal aging at temperatures of 30, 50, and 70°C (dotted line: with 4 wt.% MAPP; solid line: without MAPP) [98].

Biocomposites	Conditions	D (mm²/s)	References
Kraft fiber (40%)/PP	Water immersion at 30°C	2.90 e^{-7}	Beg and Pickering [98]
	Water immersion at 50°C	5.70 e^{-7}	
Kraft fiber (40%)/PP + 4% MAPP	Water immersion at 30°C	2.50 e^{-7}	Beg and Pickering [98, 101]
Agave fiber (30%)/PP	Water immersion	7.10 e^{-7}	Pérez-Fonseca et al. [100]
Pine sawdust (30%)/PP		4.10 e^{-7}	
Wood flour (30%)/PP	Water immersion at 27°C	5.70 e^{-4}	Mishra and Verma [96]
Kenaf fibers (30%)/PP	Water immersion	1.30 e^{-6}	Law et al. [76, 99]
Kenaf fibers (30%)/PP + 5% MAPP		8.70 e^{-7}	
Kenaf fibers (25%)/PP + 1% MAPP	Water immersion	9.40 e^{-6}	Tajvidi et al. [102]
Kenaf fibers (50%)/PP + 2% MAPP		1.30 e^{-4}	
Wood flour (25%)/PP + 1% MAPP		1.60 e^{-5}	
Wood flour (50%)/PP + 2% MAPP		1.30 e^{-5}	
Sisal fibers (30%)/PP	Water immersion	3.62 e^{-8}	Joseph et al. [97]
Diss fibers (20%)/PP	Water immersion	4.30 e^{-7}	Nouri and Tahlaiti [94]

Table 4.
Diffusion coefficients of different biocomposites reported in the literature.

pine sawdust at the same loading rate and the diffusivity was about two times higher (**Table 4**).

Furthermore, as the ambient temperature increases, the activity of the polymer molecules increases, which accelerates the diffusion of water molecules within the bio-composites. In the same line, Beg and Pickering [98] observed that increasing temperature increases the water uptake kinetics of bio-composites (**Figure 8**).

Table 4 summarizes the diffusion coefficient values of different bio-composites reported in the literature.

4.3 Hydro/hygrothermal aging

Hydro/hygrothermal aging can be qualified as a physicochemical degradation. Indeed, in the short term, the composite undergoes plasticization due to the infiltration of water molecules between the polymer chains. This type of aging is mostly reversible when the material returns to its initial state. In addition, plant fibers are known to have an affinity for water, unlike most matrices. This causes differential swelling which can also lead to micro-cracks at the fiber/matrix interphase. These cracks can propagate further through a succession of adsorption/desorption cycles (**Figure 9**). This damage is irreversible and affects the performance of bio-composites. In addition, chemical degradation of the matrix and even of the fibers can occur; this is called hydrolysis. The presence of temperature can aggravate these phenomena [103].

Law and Ishak [99] found a plasticization effect on Kenaf/PP composites after saturation, with a decrease of between 14 and 35% in all tensile and flexural mechanical properties for the composites at 40% loading. After drying the samples again, a partial recovery of their initial mechanical properties (in flexion and in traction) was noted.

Beg and Pickering [98] observed degradation in the tensile mechanical properties of kraft fiber (40%)/PP composites, with and without 4% MAPP, aged for 238 days at different temperatures of 30, 50, and 70°C. The property retention capacity was lower with increasing temperature. The authors associate these results with the degradation of the fibers and/or the fiber/matrix interface (**Figure 10**).

Freund [104] carried out a study of cyclic hygrothermal aging at 80°C on Lin/Elium composites during his thesis. Each cycle contains two phases: a saturation phase at 80%

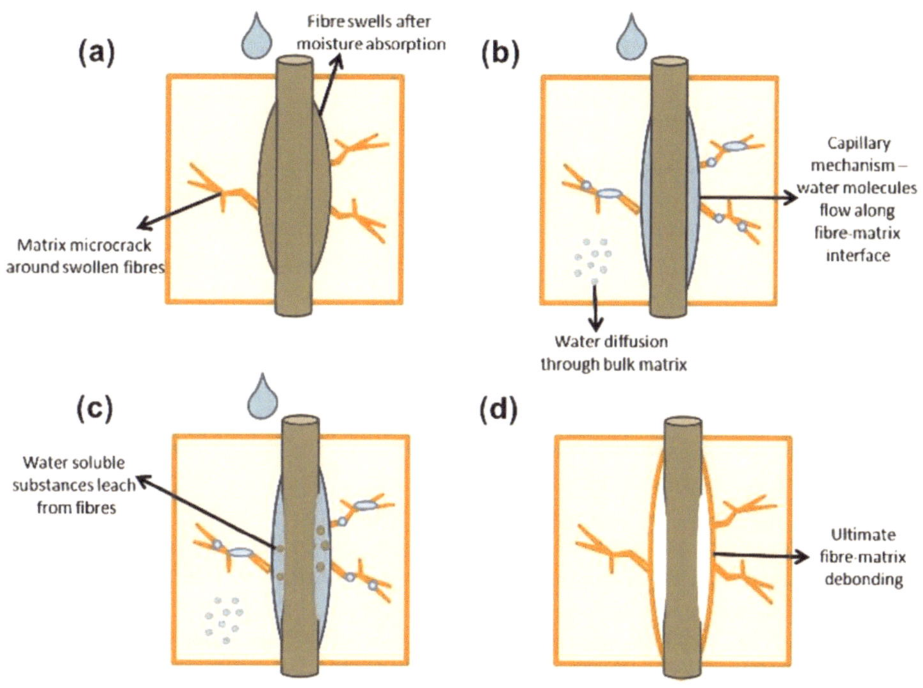

Figure 9.
Degradation mechanism during hygrothermal aging [13]: (a) fibre swells and matrix microcrack at the interface (b) water infiltration into microcracks and the interface (c) dissolution of some fibre component (d) fibres-matrix debonding.

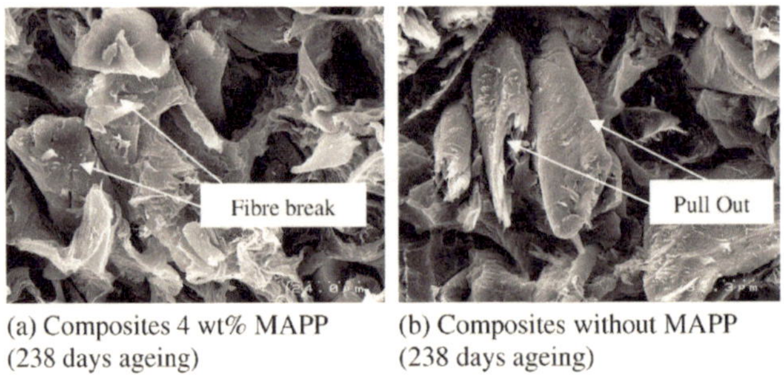

Figure 10.
Effects of hygrothermal aging on the fracture surface SEMs of composites (kraft fiber (40%)/PP) of: (a) unaged composites and (b) composites aged at 70C for 238 days [98].

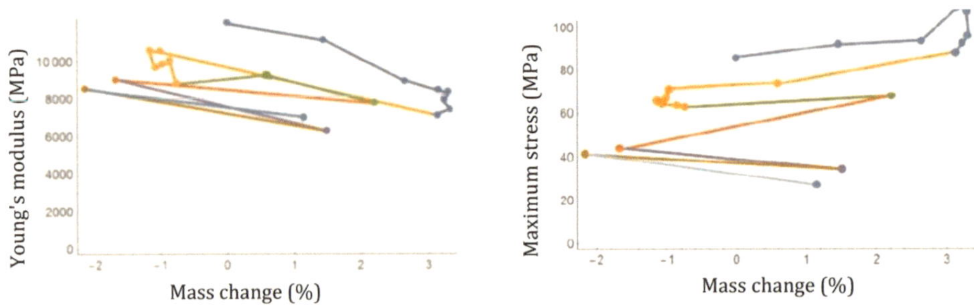

Figure 11.
Evolution of Young's modulus and maximum stress of composites flax/Elium as a function of humidity [104].

relative humidity and a drying phase at 10% relative humidity. It was found that after each cycle the composites lost mass and mechanical properties with a decrease in stiffness of about 50% after five cycles and a decrease in tensile stress from 110 to 30 MPa (**Figure 11**). These degradations also had an effect on the adsorption kinetics: the more the material undergoes aging cycles, the faster its adsorption becomes.

The author summarized this change in behavior into two reasons: swelling of the composite, especially the fibers, and degradation of the fibers. These findings were justified by a change in the failure mechanism of the composite due to a degradation of the fiber/matrix interface (**Figure 12**), and the potential degradation of the flax fibers were justified by a decrease in their crystallinity index from 77.2 to 38.4%.

On the other hand, Berges [3] studied the cyclic aging of composites at 70°C with two humidity conditions 90% RH and 15% RH for adsorption and desorption, respectively. For each cycle, a total duration of 4 days was chosen, that is, 2 days for each half cycle. Tensile tests carried out after four and nine cycles revealed a reproducible behavior for tensile modulus with a decrease in elongation at break and stress at the break with the number of cycles undergone. This reduction was related to fiber/matrix decohesion and resin damage.

Wang and Petru [105] compared the results of the mechanical properties of unidirectional (30% by volume)/epoxy flax fiber composites undergoing natural aging outdoors and artificial aging by immersion in water at 60°C. It was concluded that 60, 120, and 180 days of natural aging correspond to 1.1, 37.2, and 167.9 h of artificial aging, respectively. Comparisons were made regarding flexural strength and flexural modulus.

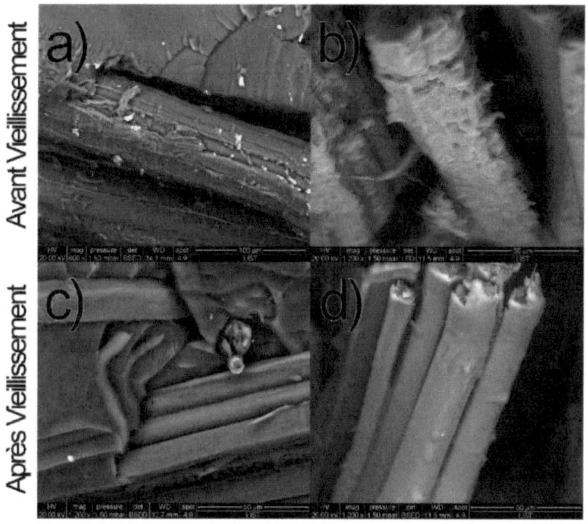

Figure 12.
Fracture facies of flax/Elium composite before and after aging [104].

5. Conclusion

Plant fibers have been shown to be highly sensitive to water molecules, which impacts the functionality of bio-composites. This phenomenon is often related to the morphology of these fibers, their cavity (lumen), and the free hydroxyl groups present on their surface. It appears that the hydrophilic character of natural fibers could be reduced by applying treatments, such as chemical and physical treatments.

The isotherm of cellulosic fibers generally has a sigmoidal shape with hysteresis loops between the adsorption and desorption curve in accordance with the type II isotherm. The Guggenheim, Anderson, and de Boer (GAB), Hailwood Horrobin (H-H), and Generalized D'Arcy and Watt (GDW) models are the most widely used in the literature to describe the isotherm of plant fibers.

The diffusive behavior of plant fibers are often based on the Fick model. However, it is sometimes described by non-Fick diffusion models. The same conclusions have been reported for bio-composites. On the other hand, the process of water diffusion through bio-composites is influenced, mainly, by two types of factors, namely internal factors (related to the bio-composite structure and the nature of its phases) and external factors (relative humidity and temperature).

Hydro/hygrothermal aging can cause irreversible damage to the bio-composite due to the different nature of the matrix and the plant fibers. This is caused by a differential swelling between these components.

Acknowledgements

This research work has been conducted with the financial support of FEDER—Region Pays de la Loire in the framework of the CIPTAP R&D project.

Author details

Nouri Mustapha and Tahlaiti Mahfoud*
Institut Catholique d'Arts et Métiers, Carquefou, France

*Address all correspondence to: mahfoud.tahlaiti@icam.fr

References

[1] Thommes M, Kaneko K, Neimark AV, Olivier JP, Rodriguez-Reinoso F, Rouquerol J, et al. Physisorption of gases, with special reference to the evaluation of surface area and pore size distribution (IUPAC Technical Report). Pure and Applied Chemistry. 2015;**87**:1051-1069. DOI: 10.1515/pac-2014-1117

[2] Sun L-M, Meunier F, Brodu N, Manero M-H. Adsorption—Aspects théoriques. 2016. p. 24

[3] Berges M. Mécanismes de dégradation sous sollicitations hydrothermomécaniques de biocomposites et renforts en fibres végétales: application au développement de mobiliers urbains ultralégers et mobiles [thesis]. Bourgogne Franche-Comté; 2018

[4] Testoni GA. In situ long-term durability analysis of biocomposites in the marine environment. n.d. p. 235

[5] Regazzi A. Contribution à l'étude du vieillissement couplé thermo-hydro-mécanique de biocomposite PLA/lin [PhD thesis]. Ecole Nationale Supérieure des Mines de Paris; 2013

[6] Fraga AN, Alvarez VA, Vazquez A, de la Osa O. Relationship between dynamic mechanical properties and water absorption of unsaturated polyester and vinyl ester glass fiber composites. Journal of Composite Materials. 2003;**37**:1553-1574. DOI: 10.1177/0021998303029421

[7] Chilali A. Étude expérimentale et modélisation de la durabilité des biocomposites à fibres de lin [thesis]. Reims; 2017

[8] Weitsman YJ, Elahi M. Effects of fluids on the deformation, strength and durability of polymeric composites—An overview. Mechanics of Time Dependent Materials. 2000;**4**:107-126. DOI: 10.1023/A:1009838128526

[9] Assarar M, Scida D, El Mahi A, Poilâne C, Ayad R. Influence of water ageing on mechanical properties and damage events of two reinforced composite materials: Flax-fibres and glass–fibres. Materials & Design. 2011; **32**:788-795. DOI: 10.1016/j.matdes.2010.07.024

[10] Roudier A. Analyse multi-échelle du comportement hygro-mécanique des fibres de lin. Génie Mécanique. Université Blaise Pascal—Clermont II; 2012

[11] Crank J. The Mathematics of Diffusion. 2nd ed. Oxford, England: Clarendon Press; 1975

[12] Gouanvé F, Marais S, Bessadok A, Langevin D, Métayer M. Kinetics of water sorption in flax and PET fibers. European Polymer Journal. 2007;**43**: 586-598. DOI: 10.1016/j.eurpolymj.2006.10.023

[13] Azwa ZN, Yousif BF, Manalo AC, Karunasena W. A review on the degradability of polymeric composites based on natural fibres. Materials and Design. 2013;**47**:424-442. DOI: 10.1016/j.matdes.2012.11.025

[14] Guo X, Wu Y. Characterizing molecular structure of water adsorbed by cellulose nanofiber film using *in situ* micro-FTIR spectroscopy. Journal of Wood Chemistry and Technology. 2017;**37**:383-392. DOI: 10.1080/02773813.2017.1306078

[15] Célino A, Gonçalves O, Jacquemin F, Fréour S. Utilisation de la spectrométrie

infrarouge pour une quantification rapide du taux d'humidité dans des fibres végétales. Revue des Composites et des Materiaux Avances. 2014;**24**: 81-95. DOI: 10.3166/rcma.24.81-95

[16] Guo X, Liu L, Wu J, Fan J, Wu Y. Qualitatively and quantitatively characterizing water adsorption of a cellulose nanofiber film using micro-FTIR spectroscopy. RSC Advances. 2018; **8**:4214-4220. DOI: 10.1039/C7RA 09894D

[17] Himmel S, Mai C. Effects of acetylation and formalization on the dynamic water vapor sorption behavior of wood. Holzforschung. 2014;**69**: 633-643. DOI: 10.1515/hf-2014-0161

[18] Stamboulis A, Baillie CA, Peijs T. Effects of environmental conditions on mechanical and physical properties of flax fibers. Composites Part A: Applied Science and Manufacturing. 2001;**32**: 1105-1115

[19] Bessadok A, Marais S, Gouanvé F, Colasse L, Zimmerlin I, Roudesli S, et al. Effect of chemical treatments of Alfa (*Stipa tenacissima*) fibres on water-sorption properties. Composites Science and Technology. 2007;**67**:685-697. DOI: 10.1016/j.compscitech.2006.04.013

[20] Bessadok A, Marais S, Roudesli S, Lixon C, Métayer M. Influence of chemical modifications on water-sorption and mechanical properties of Agave fibres. Composites Part A: Applied Science and Manufacturing. 2008;**39**:29-45. DOI: 10.1016/ j.compositesa.2007.09.007

[21] Saikia D. Studies of water absorption behavior of plant fibers at different temperatures. International Journal of Thermophysics. 2010;**31**:1020-1026. DOI: 10.1007/s10765-010-0774-0

[22] Célino A, Fréour S, Jacquemin F, Casari P. Characterization and modeling of the moisture diffusion behavior of natural fibers. Journal of Applied Polymer Science. 2013;**130**:297-306. DOI: 10.1002/app.39148

[23] Nouri M, Tahlaiti M, Grondin F, Belarbi R. The influence of chemical and thermal treatments on the Diss fiber hygroscopic behaviors. Journal of Natural Fibers. 2020:1-14. DOI: 10.1080/ 15440478.2020.1848733

[24] Kabir MM, Wang H, Lau KT, Cardona F. Chemical treatments on plant-based natural fibre reinforced polymer composites: An overview. Composites Part B: Engineering. 2012; **43**:2883-2892. DOI: 10.1016/j. compositesb.2012.04.053

[25] Alix S, Philippe E, Bessadok A, Lebrun L, Morvan C, Marais S. Effect of chemical treatments on water sorption and mechanical properties of flax fibres. Bioresource Technology. 2009;**100**: 4742-4749. DOI: 10.1016/j.biortech. 2009.04.067

[26] Kalia S, Kaith BS, Kaur I. Pretreatments of natural fibers and their application as reinforcing material in polymer composites—A review. Polymer Engineering & Science. 2009;**49**: 1253-1272. DOI: 10.1002/pen.21328

[27] Haque R, Saxena M, Shit SC, Asokan P. Fibre-matrix adhesion and properties evaluation of sisal polymer composite. Fibers and Polymers. 2015; **16**:146-152. DOI: 10.1007/s12221-015-0146-2

[28] Alao PF, Marrot L, Kallakas H, Just A, Poltimäe T, Kers J. Effect of hemp fiber surface treatment on the moisture/ water resistance and reaction to fire of reinforced PLA composites. Materials. 2021;**14**:4332. DOI: 10.3390/ma14154332

[29] Balaji AN, Nagarajan KJ. Characterization of alkali treated and untreated new cellulosic fiber from Saharan *Aloe vera* cactus leaves. Carbohydrate Polymers. 2017;**174**: 200-208. DOI: 10.1016/j.carbpol.2017.06.065

[30] Asim M, Jawaid M, Abdan K, Ishak MR. Effect of alkali and silane treatments on mechanical and fibre-matrix bond strength of Kenaf and pineapple leaf fibres. Journal of Bionic Engineering. 2016;**13**:426-435. DOI: 10.1016/S1672-6529(16)60315-3

[31] Gassan J, Gutowski VS. Effects of corona discharge and UV treatment on the properties of jute-fibre epoxy composites. Composites Science and Technology. 2000;**60**:2857-2863

[32] Pizzi A, Kueny R, Lecoanet F, Massetau B, Carpentier D, Krebs A, et al. High resin content natural matrix–natural fibre biocomposites. Industrial Crops and Products. 2009;**30**:235-240. DOI: 10.1016/j.indcrop.2009.03.013

[33] Ragoubi M, Bienaimé D, Molina S, George B, Merlin A. Impact of corona treated hemp fibres onto mechanical properties of polypropylene composites made thereof. Industrial Crops and Products. 2010;**31**:344-349. DOI: 10.1016/j.indcrop.2009.12.004

[34] Sinha E, Panigrahi S. Effect of plasma treatment on structure, wettability of jute fiber and flexural strength of its composite. Journal of Composite Materials. 2009;**43**: 1791-1802. DOI: 10.1177/002199 8309338078

[35] Graupner N, Albrecht K, Hegemann D, Müssig J. Plasma modification of man-made cellulose fibers (Lyocell) for improved fiber/ matrix adhesion in poly(lactic acid) composites. Journal of Applied Polymer Science. 2013;**128**:4378-4386. DOI: 10.1002/app.38663

[36] Marais S, Gouanvé F, Bonnesoeur A, Grenet J, Poncin-Epaillard F, Morvan C, et al. Unsaturated polyester composites reinforced with flax fibers: Effect of cold plasma and autoclave treatments on mechanical and permeation properties. Composites Part A: Applied Science and Manufacturing. 2005;**36**:975-986. DOI: 10.1016/j.compositesa.2004.11.008

[37] Ariawan D, Salim MS, Taib RM, Thirmizir MZA, Ishak ZAM. Interfacial characterisation and mechanical properties of heat treated non-woven Kenaf fibre and its reinforced composites. Composite Interfaces. 2018; **25**:187-203. DOI: 10.1080/09276440. 2017.1354562

[38] Rong MZ, Zhang MQ, Liu Y, Yang GC, Zeng HM. The effect of fiber treatment on the mechanical properties of unidirectional sisal-reinforced epoxy composites. Composites Science and Technology. 2001;**61**:1437-1447. DOI: 10.1016/S0266-3538(01)00046-X

[39] Fook LT, Yatim JM. An experimental study on the effect of alkali treatment on properties of Kenaf fiber for reinforced concrete elements. International Journal of Research in Engineering and Technology. 2015;**4**: 37-40. DOI: 10.15623/ijret.2015. 0408007s

[40] Chandrasekar M, Ishak MR, Sapuan SM, Leman Z, Jawaid M. A review on the characterisation of natural fibres and their composites after alkali treatment and water absorption. Plastics, Rubber and Composites. 2017;**46**: 119-136. DOI: 10.1080/14658011.2017. 1298550

[41] Huang Z, Ge H, Yin J, Liu F. Effects of fiber loading and chemical treatments

on properties of sisal fiber-reinforced sheet molding compounds. Journal of Composite Materials. 2017;**51**:3175-3185. DOI: 10.1177/0021998316683440

[42] Faruk O, Bledzki AK, Fink H-P, Sain M. Biocomposites reinforced with natural fibers: 2000–2010. Progress in Polymer Science. 2012;**37**:1552-1596. DOI: 10.1016/j.progpolymsci.2012.04.003

[43] Bledzki AK, Mamun AA, Lucka-Gabor M, Gutowski VS. The effects of acetylation on properties of flax fibre and its polypropylene composites. Express Polymer Letters. 2008;**2**:413-422. DOI: 10.3144/expresspolymlett.2008.50

[44] Bledzki A. Composites reinforced with cellulose based fibres. Progress in Polymer Science. 1999;**24**:221-274. DOI: 10.1016/S0079-6700(98)00018-5

[45] Vishnu Vardhini K, Murugan R, Surjit R. Effect of alkali and enzymatic treatments of banana fibre on properties of banana/polypropylene composites. Journal of Industrial Textiles. 2018;**47**:1849-1864. DOI: 10.1177/1528083717714479

[46] Li X, Tabil LG, Panigrahi S. Chemical treatments of natural fiber for use in natural fiber-reinforced composites: A review. Journal of Polymers and the Environment. 2007;**15**:25-33. DOI: 10.1007/s10924-006-0042-3

[47] Motaung TE, Linganiso LZ, Kumar R, Anandjiwala RD. Agave and sisal fibre inforced polyfrurfuryl alcohol composite. Journal of Thermoplastic Composite Materials. 2017;**30**:1323-1343. DOI: 10.1177/0892705716632858

[48] Bledzki AK, Reihmane S, Gassan J. Properties and modification methods for vegetable fibers for natural fiber composites. Journal of Applied Polymer Science. 1996;**59**:1329-1336. DOI: 10.1002/(SICI)1097-4628(19960222)59:8<1329::AID-APP17>3.0.CO;2-0

[49] Placet V, Day A, Beaugrand J. The influence of unintended field retting on the physicochemical and mechanical properties of industrial hemp bast fibres. Journal of Materials Science. 2017;**52**:5759-5777. DOI: 10.1007/s10853-017-0811-5

[50] Salmén L, Larsson PA. On the origin of sorption hysteresis in cellulosic materials. Carbohydrate Polymers. 2018;**182**:15-20. DOI: 10.1016/j.carbpol.2017.11.005

[51] Shi J, Avramidis S. Water sorption hysteresis in wood: I review and experimental patterns—Geometric characteristics of scanning curves. Holzforschung. 2017;**71**:307-316. DOI: 10.1515/hf-2016-0120

[52] Fredriksson M, Thybring EE. Scanning or desorption isotherms? Characterising sorption hysteresis of wood. Cellulose. 2018;**25**:4477-4485. DOI: 10.1007/s10570-018-1898-9

[53] Thygesen L, Thybring E, Hoffmeyer P. Water sorption in wood and modified wood at high values of relative humidity. Part I: Results for untreated, acetylated, and furfurylated Norway spruce. Holzforschung. 2010;**64**:315-323. DOI: 10.1515/HF.2010.044

[54] Keating BA, Hill CAS, Sun D, English R, Davies P, McCue C. The water vapor sorption behavior of a galactomannan cellulose nanocomposite film analyzed using parallel exponential kinetics and the Kelvin-Voigt viscoelastic model. Journal of Applied Polymer Science. 2013;**129**:2352-2359. DOI: 10.1002/app.39132

[55] Hill CAS, Norton A, Newman G. The water vapor sorption behavior of natural fibers. Journal of Applied Polymer Science. 2009;**112**:1524-1537. DOI: 10.1002/app.29725

[56] Trabelsi A, Belarbi R, Turcry P, Aït-Mokhtar A. Characterization of water vapor desorption variability and its effects on drying simulation for an in-situ high performances concrete. Construction and Building Materials. 2011;**141**:140–151

[57] Bessadok A, Langevin D, Gouanvé F, Chappey C, Roudesli S, Marais S. Study of water sorption on modified Agave fibres. Carbohydrate Polymers. 2009;**76**: 74-85. DOI: 10.1016/j.carbpol.2008. 09.033

[58] Marzec A, Lewicki PP. Antiplasticization of cereal-based products by water. Part I. Extruded flat bread. Journal of Food Engineering. 2006;**73**:1-8. DOI: 10.1016/j.jfoodeng. 2004.12.002

[59] Koko CA, Diomande M, Kouame BK, Assidjo EN. Détermination expérimentale et modélisation des isothermes d'adsorption d'eau des amandes d'Irvingia gabonensis de la région du Haut-Sassandra (Côte d'Ivoire). n.d. p. 17

[60] Bui R, Labat M, Aubert J-E. Comparison of the saturated salt solution and the dynamic vapor sorption techniques based on the measured sorption isotherm of barley straw. Construction and Building Materials. 2017;**141**:140-151. DOI: 10.1016/ j.conbuildmat.2017.03.005

[61] Patcharawijit A, Choodum N, Yamsaengsung R. Effects of superheated steam treatment on moisture adsorption and mechanical properties of pre-dried rubberwood. Drying Technology. 2019; **37**:1647-1655. DOI: 10.1080/07373937. 2018.1529037

[62] Furmaniak S, Terzyk AP, Gauden PA, Rychlicki G. Applicability of the generalised D'Arcy and Watt model to description of water sorption on pineapple and other foodstuffs. Journal of Food Engineering. 2007;**79**:718-723. DOI: 10.1016/j.jfoodeng.2006.02.036

[63] Mannan KM, Talukder MAI. Characterization of raw, delignified and bleached jute fibres by study of absorption of moisture and some mechanical properties. Polymer. 1997;**38**: 2493-2500. DOI: 10.1016/S0032-3861 (96)00804-X

[64] Loh WK, Crocombe AD, Abdel Wahab MM, Ashcroft IA. Modelling anomalous moisture uptake, swelling and thermal characteristics of a rubber toughened epoxy adhesive. International Journal of Adhesion and Adhesives. 2005;**25**:1-12. DOI: 10.1016/j.ijadhadh. 2004.02.002

[65] Carter HG, Kibler KG. Langmuir-type model for anomalous moisture diffusion in composite resins. Journal of Composite Materials. 1978;**12**:118-131. DOI: 10.1177/002199837801200201

[66] Alves Fidelis ME, Pereira TVC, Gomes O da FM, de Andrade Silva F, Toledo Filho RD. The effect of fiber morphology on the tensile strength of natural fibers. Journal of Materials Research and Technology. 2013;**2**: 149-157. DOI: 10.1016/j.jmrt.2013. 02.003

[67] Ferreira SR, Pepe M, Martinelli E, de Andrade Silva F, Toledo Filho RD. Influence of natural fibers characteristics on the interface mechanics with cement based matrices. Composites Part B: Engineering. 2018;**140**:183-196. DOI: 10.1016/j.compositesb.2017.12.016

[68] Newman RH, Le Guen MJ, Battley MA, Carpenter JEP. Failure mechanisms in composites reinforced with unidirectional Phormium leaf fibre. Composites Part A: Applied Science and Manufacturing. 2010;**41**:353-359. DOI: 10.1016/j.compositesa.2009.11.001

[69] Nouri M, Griballah I, Tahlaiti M, Grondin F, Beaugrand J. Plant extraction and physicochemical characterizations of untreated and pretreated Diss fibers (*Ampelodesmos mauritanicus*). Journal of Natural Fibers. 2019:1-11. DOI: 10.1080/15440478.2019.1687062

[70] Nouri M, Tahlaiti M, Meroufel A, Grondin F. Enhanced modeling of water diffusion in natural fibers: Application to Diss fibers. Journal of Natural Fibers. 2021:1-10. DOI: 10.1080/15440478.2021.1982817

[71] Wang W, Guo X, Zhao D, Liu L, Zhang R, Yu J. Water absorption and hygrothermal aging behavior of wood-polypropylene composites. Polymers. 2020;**12**:782. DOI: 10.3390/polym12040782

[72] Chandekar H, Chaudhari V, Waigaonkar S, Mascarenhas A. Effect of chemical treatment on mechanical properties and water diffusion characteristics of jute-polypropylene composites. Polymer Composites. 2019; **41**:1447-1461. DOI: 10.1002/pc.25468

[73] Ameer MH, Nawab Y, Ali Z, Imad A, Ahmad S. Development and characterization of jute/polypropylene composite by using comingled nonwoven structures. Journal of the Textile Institute. 2019;**110**:1652-1659. DOI: 10.1080/00405000.2019.1612502

[74] Boukettaya S, Alawar A, Almaskari F, Ben Daly H, Abdala A, Chatti S. Modeling of water diffusion mechanism in polypropylene/date palm fiber composite materials. Journal of Composite Materials. 2018;**52**:2651-2659. DOI: 10.1177/0021998317752228

[75] Kord B, Hosseinihashemi SK, Modirzare M. Influence of fungal infection on the long-term water absorption and morphological behavior of bagasse fiber/polypropylene composites at different exposure times. Science and Engineering of Composite Materials. 2013;**20**:351-357. DOI: 10.1515/secm-2013-0005

[76] Lee JM, Mohd Ishak ZA, Mat Taib R, Law TT, Ahmad Thirmizir MZ. Mechanical, thermal and water absorption properties of kenaf-fiber-based polypropylene and poly(butylene succinate) composites. Journal of Polymers and the Environment. 2013;**21**: 293-302. DOI: 10.1007/s10924-012-0516-4

[77] Hosseinihashemi SK, Arwinfar F, Najafi A, Nemli G, Ayrilmis N. Long-term water absorption behavior of thermoplastic composites produced with thermally treated wood. Measurement. 2016;**86**:202-208. DOI: 10.1016/j.measurement.2016.02.058

[78] Gozdecki C, Wilczyn'ski A. Effects of wood particle size and test specimen size on mechanical and water resistance properties of injected wood–high density polyethylene composite. Wood and Fiber Science. 2015;**47**:365-374

[79] Mrad H, Alix S, Migneault S, Koubaa A, Perré P. Numerical and experimental assessment of water absorption of wood-polymer composites. Measurement. 2018;**115**:197-203. DOI: 10.1016/j.measurement.2017.10.011

[80] Jiang N, Li Y, Li D, Yu T, Li Y, Xu J, et al. 3D finite element modeling of water diffusion behavior of jute/PLA composite based on X-ray computed

tomography. Composites Science and Technology. 2020;**199**:108313. DOI: 10.1016/j.compscitech.2020.108313

[81] Sanjeevi S, Shanmugam V, Kumar S, Ganesan V, Sas G, Johnson DJ, et al. Effects of water absorption on the mechanical properties of hybrid natural fibre/phenol formaldehyde composites. Scientific Reports. 2021;**11**:13385. DOI: 10.1038/s41598-021-92457-9

[82] Shen C-H, Springer GS. Moisture absorption and desorption of composite materials. Journal of Composite Materials. 1976;**10**:2-20. DOI: 10.1177/002199837601000101

[83] Rangaraj SV, Smith LV. Effects of moisture on the durability of a wood/thermoplastic composite. Journal of Thermoplastic Composite Materials. 2000;**13**:140-161. DOI: 10.1177/089270570001300204

[84] Cheour K, Assarar M, Scida D, Ayad R, Gong X-L. Effect of water ageing on the mechanical and damping properties of flax-fibre reinforced composite materials. Composite Structures. 2016;**152**:259-266. DOI: 10.1016/j.compstruct.2016.05.045

[85] Sahu P, Gupta MK. Lowering in water absorption capacity and mechanical degradation of sisal/epoxy composite by sodium bicarbonate treatment and PLA coating. Polymer Composites. 2020;**41**:668-681. DOI: 10.1002/pc.25397

[86] Kumosa L, Benedikt B, Armentrout D, Kumosa M. Moisture absorption properties of unidirectional glass/polymer composites used in composite (non-ceramic) insulators. Composites Part A: Applied Science and Manufacturing. 2004;**35**:1049-1063. DOI: 10.1016/j.compositesa.2004.03.008

[87] Jiang X, Kolstein H, Bijlaard FSK. Moisture diffusion in glass–fiber-reinforced polymer composite bridge under hot/wet environment. Composites Part B: Engineering. 2013;**45**:407-416. DOI: 10.1016/j.compositesb.2012.04.067

[88] Bhatti IS, Yan Z. Comparative hygrothermal ageing analysis of glass and carbon fiber polymer composites: Numerical and experimental study. In: 2015 12th International Bhurban Conference on Applied Sciences and Technology (IBCAST). 2015. pp. 28-33. DOI:10.1109/IBCAST.2015.7058474

[89] Joliff Y, Belec L, Chailan JF. Modified water diffusion kinetics in an unidirectional glass/fibre composite due to the interphase area: Experimental, analytical and numerical approach. Composite Structures. 2013;**97**:296-303. DOI: 10.1016/j.compstruct.2012.09.044

[90] Joliff Y, Belec L, Heman MB, Chailan JF. Experimental, analytical and numerical study of water diffusion in unidirectional composite materials—Interphase impact. Computational Materials Science. 2012;**64**:141-145. DOI: 10.1016/j.commatsci.2012.05.029

[91] Jain D, Mukherjee A. Three-dimensional hygromechanical analysis of fibre polymer composites: Effect of boundary conditions. Composites Part B: Engineering. 2016;**90**:203-211. DOI: 10.1016/j.compositesb.2015.12.011

[92] Peret T, Clement A, Freour S, Jacquemin F. Numerical transient hygro-elastic analyses of reinforced Fickian and non-Fickian polymers. Composite Structures. 2014;**116**:395-403. DOI: 10.1016/j.compstruct.2014.05.026

[93] Bond DA. Moisture diffusion in a fiber-reinforced composite: Part I—Non-Fickian transport and the effect of fiber spatial distribution. Journal of

Composite Materials. 2005;**39**:2113-2141. DOI: 10.1177/0021998305052030

[94] Nouri M, Tahlaiti M. A dual-scale numerical model for the diffusive behaviour prediction of biocomposites based on randomly oriented fibres. Construction Technologies and Architecture. 2022;**1**:584-587. DOI: 10.4028/www.scientific.net/CTA.1.584

[95] Masoodi R, Pillai KM. A study on moisture absorption and swelling in bio-based jute-epoxy composites. Journal of Reinforced Plastics and Composites. 2012;**31**:285-294. DOI: 10.1177/0731684411434654

[96] Mishra S, Verma J. Effect of compatibilizers on water absorption kinetics of polypropylene/wood flour foamed composites. Journal of Applied Polymer Science. 2006;**101**:2530-2537. DOI: 10.1002/app.23916

[97] Joseph PV, Rabello MS, Mattoso LHC, Joseph K, Thomas S. Environmental effects on the degradation behaviour of sisal fibre reinforced polypropylene composites. Composites Science and Technology. 2002;**62**:1357-1372. DOI: 10.1016/S0266-3538(02)00080-5

[98] Beg MDH, Pickering KL. Mechanical performance of Kraft fibre reinforced polypropylene composites: Influence of fibre length, fibre beating and hygrothermal ageing. Composites Part A: Applied Science and Manufacturing. 2008;**39**:1748-1755. DOI: 10.1016/j.compositesa.2008.08.003

[99] Law TT, Ishak ZAM. Water absorption and dimensional stability of short Kenaf fiber-filled polypropylene composites treated with maleated polypropylene. Journal of Applied Polymer Science. 2011;**120**:563-572. DOI: 10.1002/app.33184

[100] Pérez-Fonseca AA, Robledo-Ortíz JR, Moscoso-Sánchez FJ, Rodrigue D, González-Núñez R. Injection molded self-hybrid composites based on polypropylene and natural fibers. Polymer Composites. 2014;**35**:1798-1806. DOI: 10.1002/pc.22834

[101] Beg MDH, Pickering KL. Reprocessing of wood fibre reinforced polypropylene composites. Part II: Hygrothermal ageing and its effects. Composites Part A: Applied Science and Manufacturing. 2008;**39**:1565-1571. DOI: 10.1016/j.compositesa.2008.06.002

[102] Tajvidi M, Najafi SK, Moteei N. Long-term water uptake behavior of natural fiber/polypropylene composites. Journal of Applied Polymer Science. 2006;**99**:2199-2203. DOI: 10.1002/app.21892

[103] Bergeret A, Krawczak P. Liaison renfort/matrice—Définition et caractérisation. Ref: TIP100WEB—“Plastiques et composites”. 2006. Available from: https://www.techniques-ingenieur.fr/base-documentaire/materiaux-th11/caracterisation-et-proprietes-d-usage-des-composites-42144210/liaison-renfort-matrice-am5305/ [Accessed: 19 July 2020]

[104] Freund L. Étude du vieillissement hygrothermique des composites renforcés de fibres naturelles: Approche expérimentale et modélisation. n.d. p. 165

[105] Wang X, Petrů M. Degradation of bending properties of flax fiber reinforced polymer after natural aging and accelerated aging. Construction and Building Materials. 2020;**240**:117909. DOI: 10.1016/j.conbuildmat.2019.117909

Chapter 5

Non-destructive Characterizations of Natural Yarns and Fabrics

Ruksana Baby, Kavita Mathur and Emiel DenHartog

Abstract

Textiles, next to skin, are an integral part of our lives, govern the skin microclimate, and contribute to our comfort and health. Over the years, natural and synthetic textiles have dominated the industry in diverse application areas. However, when it comes to the sustainability of the raw materials or products, processes, and the environment, the natural polymers or fibers will always dominate the preference. One of the many natural fibers, cotton fiber is the most popular and widely used one, leading to many fundamental researches in the fields of polymers, fibers, fabrics, their manufacturing processes and finishing, as well as in product characterizations and performance evaluations. To-date, most textile-characterization techniques involve processes which compromise the morphology of the textiles being tested, and are mostly destructive. In this chapter, a few novel non-destructive characterizations of textiles, made from natural fibers (specifically cotton), will be discussed which involve X-ray micro-computed tomographic (XRM-CT) three-dimensional (3D) image analysis. Tomographic characterizations allow the investigation of both the surface profiles and the inner construction of the textiles without compromising the morphology. The findings discussed in this chapter will assist in non-destructive characterizations and performance evaluations of other diverse material classes as well.

Keywords: woven fabric, cotton, XRM-CT, fabric morphology, skin comfort

1. Introduction

Textiles are complex porous structures, composed of fibers, yarns and fabrics. It refers to a vast and diverse application areas, and are an integral part of our lives used in many routine applications such as apparel, shoes, accessories and home furnishings. The functionality and performance of different textile products depends on their physical, mechanical, and thermal properties, as well as, air and moisture (and/or liquid) transport properties. Hence, in order to optimize and improve the performance of the textiles in diverse application areas, characterization of the textiles and an understanding of their structure–property relationships are imperative. For instance, textiles next to skin regulate the skin micro-climate, and skin-textile friction (measured by friction force or friction coefficient) play a vital role in skin comfort and health in different conditions [1]. The frictional properties of textiles can become an important issue when it expedites the development of skin injuries. For example, excessive and repetitive friction from moist textiles (due to absorption of skin sweat) were reported to expedite tissue deformation and skin damage, friction blisters, pressure ulcers (also known as decubitus ulcers), and even more severe unwanted problems in athletes, military, and in people with compromised skin conditions and immobility [1–11]. Therefore, to optimize the fabric frictional properties, depending on applications, it is essential to have an understanding of the structural characteristics of the fibers, yarns and fabrics, and their effects on respective physical and mechanical properties in different micro-climatic conditions. These understanding can provide valuable insights on how to engineer and design textiles to lower the friction force when in contact with the skin [12, 13].

Textile characterization is also a broad topic which involves the study of polymers, fibers, yarns, and fabrics as well as their end products in diverse applications. Over the years, there has been a significant number of studies by scientists to develop theories and systematic test methods, and standardize these methods to ensure consistency in textile characterization (methods) all around the world. However, most characterization methods are destructive in nature, and/or compromise the structure during tests which may compromise the resultant data. For instance, measurement of yarn crimp% in fabrics following the ASTM D3883 (option A) test method involves the extraction of the yarns from the fabrics which may influence the measurements by releasing the tension acting on the yarns [12, 14]. In addition, natural fibers (for example, cotton) tend to be hairy on the surface. The hairiness on the fabric surfaces makes the characterization of natural fibers, yarns and fabrics more challenging. This is why there is a growing interest in nondestructive characterization of textiles. Literature reveals the use of the X-ray micro-computed tomography (XRM-CT or X-ray CT) systems for nondestructive characterization of fibers [15–17], yarns [12, 18] and fabrics [12, 19–22]. The X-ray CT system has also been well-known for nondestructive analysis and its diverse applications in biological and medical science [23], material science [24], and in the analysis of membranes [25]. The nondestructive analysis reported in these studies [12, 15–25] involved acquiring high resolution 3D images of the test samples using the X-ray CT system. CT images are reconstructed and then imported to processing software for advanced analysis of both the 2D and 3D images [12]. CT images can also be analyzed to represent different fibers with different densities using gradients of gray levels or intensities. In this chapter, a few non-destructive yarn and fabric characterization methods using the X-ray CT system and image analysis tools will be discussed. In addition, the characterized properties will be compared with those obtained from the existing classical test standards. It is to be noted that the chapter will focus on the interpretation of the test results of cotton-made yarns and fabrics.

2. Yarn and fabric characterizations

Yarn and fabric properties can be measured using theoretical calculation and classical test standards as well as the X-ray micro-computed tomographic 3D images of the fabrics. This section will discuss the measurement methods in details. The fabrics which will be used for comparison included four plain and four satin woven fabrics made from 100% cotton ring-spun yarns (40 and 60Ne linear density) [12, 13]. The fabrics varied in thread density, and were grouped into 'low density' and 'high density' fabrics for respective weave designs and yarn counts. **Table 1** below represents the fabrics.

Sample code	Yarn linear density (Ne)	Weave design	Fabric density
P40L	40	Plain	Low
P40H	40	Plain	High
S40L	40	5-H Satin	Low
S40H	40	5-H Satin	High
P60L	60	Plain	Low
P60H	60	Plain	High
S60L	60	5-H Satin	Low
S60H	60	5-H Satin	High

Table 1.
Woven fabrics were coded for easy identification where the 1st letter indicated weave design (P = plain and S = satin), the middle two-digit numbers indicated yarn count or linear density (40 for 40Ne, and 60 for 60Ne), and the 4th letter indicate fabric thread density (L = low and H = high). For example, P40L stands for low-density plain-woven fabric made of 40Ne yarn [12, 13].

2.1 Theoretical evaluation and classical test methods

Yarn Diameter: Given the yarn count (yarn linear density), yarn diameter can be theoretically calculated using the Eq. (1) below [12, 13, 26]. In the development of the equation, the yarns were assumed to be uniform round cylinders and the diameters were expressed in terms of yarn count (tex system).

$$d = \frac{1}{280.2}\sqrt{\frac{N_t}{\varnothing \rho_f}} \tag{1}$$

Specification and Unit:
d = yarn diameter (cm).
N_t = yarn count (tex, expressed in g/km);
14.76 tex (for 40Ne), and 9.84 tex (for 60Ne)
$\varnothing$ = yarn packing factor (constant, and varies with spinning type).
0.6 For ring-spun yarns
ρ_f = fiber density (g/cm^3); 1.52 g/cm^3 for cotton fiber.

For example, the calculated yarn diameters were 0.014 and 0.012 cm for 40 and 60Ne yarns respectively. The diameter units were then converted to other units (such as to mm) as needed for further analysis [12, 13].

Yarn Packing Factor or Packing Density: Yarn packing factor can be theoretically calculated using the Eq. (2) below [12, 13, 27, 28].

$$\text{Packing factor} = \frac{\text{Fiber Area}}{\text{Yarn Area}} = \frac{\frac{\text{Yarn count (in denier)}}{\text{Volumetric density of fiber}, \rho_f}}{\text{Yarn area}, \ (\pi r^2)} \tag{2}$$

The calculated counts (in denier) for 40 and 60Ne yarns were 132.87 denier and 88.58 denier respectively, and yarn area was calculated using the theoretical yarn diameters discussed earlier.

Yarn Twists: Yarn twists can be measured by untwist-retwist method (revolution/ 20 inch) following the ASTM D1422–99(2008) test standard which satisfactorily determines the approximate twist in all ring-spun yarns and 100% cotton open-end spun yarns [29]. **Figure 1**. shows a RU-493 power-driven Twist Tester. In this method, 25 yarns are extracted from each fabric and the twist direction is identified. One end of a 10″ yarn sample is pulled through the stationary clamp (then immediately closed), and the other end is fastened to the rotational clamp (**Figure 1**) ensuring the marker (or pointer) is in the zero position. Twist direction is then selected on the tester to let the untwisting/rotation of yarn in the same twist direction. The motor is turned on, allowing yarns to untwist and elongate by releasing the yarn tension, and then retwist back in the opposite direction. When the marker returns to zero position, the motor is stopped. Number of turns is recorded from the reading dial and yarn twists (in TPI value that stands for Twists per Inch) can be calculated using Eq. (3). In the equation, the value 2 implies untwisting and retwisting. Twist multiplier (TM) can be then calculated using Eq. (4), and the entire process is repeated for the remaining 24 yarn samples extracted from fabrics [27–29].

$$\mathrm{TPI} = \frac{R}{2L} \tag{3}$$

$$\mathrm{TM} = \frac{\mathrm{TPI}}{\sqrt{\mathrm{Ne}}} \tag{4}$$

Yarn Crimp: ASTM D3883, 'Standard Test Method for Yarn Crimp and Yarn Take-Up in Woven Fabrics' is usually followed for yarn crimp measurements [14]. In both warp and weft directions on fabrics, marks are made 10″ apart, and yarns are unraveled for measurements. Following option, A of ASTM D3883, a yarn is stretched to the point of no crimp, and the distance between yarn markings in its stretched state is measured. The process is repeated 10 times for both warp (also known as ends) and weft (also known as picks or filling yarns) yarns of each fabric.

$$\text{Yarn Crimp\%} = 100 \times \frac{(\text{Straightened yarn length} - \text{marked yarn length in fabric})}{\text{marked yarn length in fabric}} \tag{5}$$

Yarn crimp (%) can be then calculated using Eq. (5). Note 2 of ASTM D3883 states that Option A may lead to variation from the possible inconsistency during stretching of yarns by hand since the stretch force is unknown.

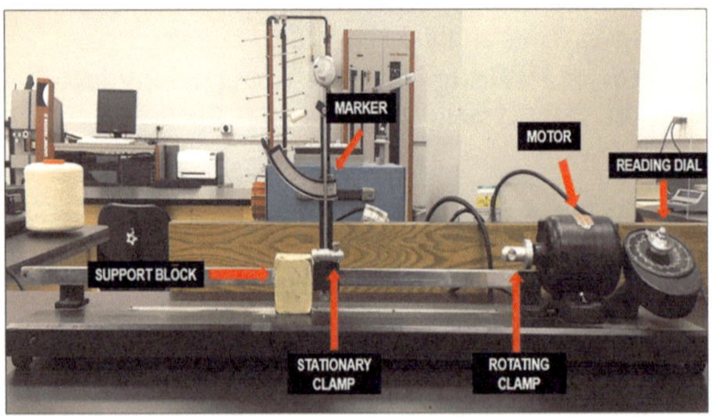

Figure 1.
RU-493 power-driven twist tester for yarn twists measurements [27].

Fabric Thread Density: Fabric thread density are determined following the ASTM 3775, Standard Test Method for End (warp) and Pick (weft/filling) Count of Woven Fabrics [30]. 1-inch marks are made on fabrics in both the warp and weft directions. Fabrics are then cut cautiously near a mark to unravel before the mark. Using a handheld pick, yarns are pulled out of the fabric while counting, and the total number of yarns within the 1-inch marks is reported as ends/inch (EPI) and picks/inch (PPI).

Fabric Basis Weight: Basis weight of the fabrics (g/m^2or gsm) can be measured following the ASTM D3776, standard test methods for mass per unit area (weight) of fabric [31]. Samples are cut into 6″ × 6″ size, and weighed (g) in a high precision electronic balance. The values are then recorded in g/m^2 and the process is repeated 5 times for each fabric.

Fabric Thickness: Thickness of the fabrics can be measured in a thickness gauge (Ames Digital Comparator; model #3-P1500; displayed in **Figure 2**) following the ASTM D1777–96(2002) Standard Test Method for Thickness of Textile Materials (table option 1; 4.14 ± 0.21 KPa) [12, 13, 27, 28, 32]. A sample is placed (technical face up) on the base of the instrument called anvil, and a weighted pressure foot is lowered. Pressure is applied for at least 5–6 s. Thickness (mm) is determined by the distance between anvil and pressure foot and recorded. The process is repeated at least 10 times for each fabric.

Fabric Cover Factor: Fabric Cover factor is expressed as percentage (%), and refers to the area of the fabric which is actually covered by fibers and yarns. It is defined as the ratio of surface area covered by yarns to total fabric surface area. Fabric cover factor can be calculated using Eqs. (6)–(8) [12, 13].

$$C_f = (C_1 + C_2 - C_1C_2) \times 100 \tag{6}$$

$$C_1 = p_1 \times d_1 \tag{7}$$

$$C_2 = p_2 \times d_2 \tag{8}$$

Specification and Unit:
C_f = fabric cover factor (%).
C_1 = warp cover factor.
C_2 = weft cover factor.

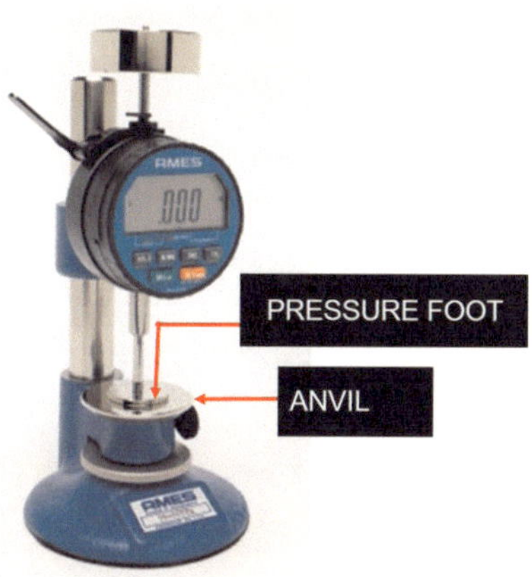

Figure 2.
Ames digital thickness gauge (model #3-P1500) [27].

p_1, and p_2 = fabric thread density; warp or ends per inch (EPI), and weft or picks per inch (PPI) respectively (measured following ASTM 3775).

d_1, and d_2 = warp and weft yarn diameters (inch) respectively (theoretically calculated).

2.2 X-ray micro-computed tomographic 3D image analysis

Recent research [12] demonstrated the use of an Xradia 510 Versa 3D X-ray microscope (XRM) (Zeiss, Germany) (**Figure 3**) for imaging the fabric samples. Fabrics were mounted on the sample holder maintaining warp in the vertical direction to ensure accuracy and consistency in the test method. The high-resolution images (pixel size: 1.31 μm) were obtained at 50 kV and 10 W using the 4X objective lens, and from a projection number set to 1601. The source-to-fabric and detector-to-fabric distances were maintained constant for all fabrics to ensure same resolution at the same magnification. The images were then imported to the XMReconstructor software for post-reconstruction into 8-bit TIFF files with a size of 980×1008×990 (width, height, and depth) [12, 13].

The reconstructed TIFF images were then imported into the Dragonfly Pro software (ORS, Montreal, Canada). Window leveling, contrast and intensity space were adjusted for both the 3D (**Figure 4a**) and 2D images (**Figure 4b**), and image segments were created for a more meaningful visualization as well as for different sets of measurements. New segmented regions of interest (ROIs) were created which highlighted the fibers (**Figure 4c**), noise was removed (**Figure 4d**), and then the images were converted to binary scale (**Figure 4e**) for analysis. The 2D planes were then adjusted to be as perpendicular to each other as possible (**Figure 5**) where *X*, *Y* and *Z*-axis indicated thickness, warp and weft directions respectively. In the images, the black color represented air while the white represented the fibers in the fabrics [12, 13].

Yarn Diameter: Yarn diameters (μm) were measured from the 2D images using the scales in the Dragonfly Pro software (**Figure 6a**). Warp and weft yarn diameters were measured using the 2D XZ and XY images respectively (as depicted in **Figure 5**). Fifty measurements (μm) were taken and averaged for yarn diameters for both warp and weft yarns, and then converted into cm [12, 13].

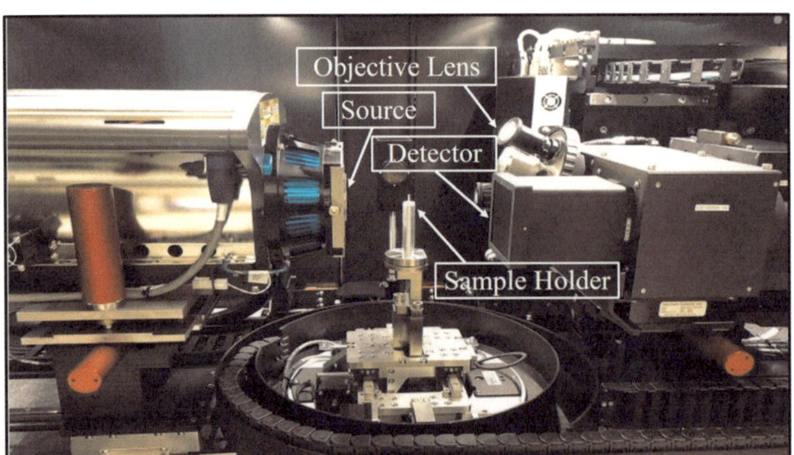

Figure 3.
Zeiss Xradia 510 versa 3D X-ray tomography system (inside view) [12, 13].

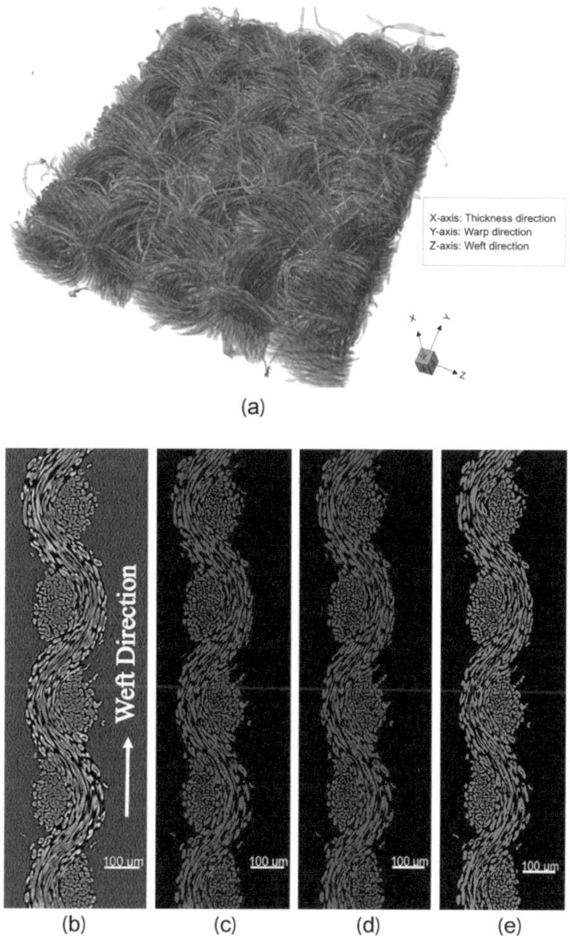

Figure 4.
3D and 2D views of P40H fabric, obtained from Xradia 510 versa 3D X-ray microscope (XRM): (a) 3D image, (b) 2D view, original image, (c) 2D view of image segment with noise, (d) 2D view of image segment after noise removal, and (e) 2D binary image of the segment highlighting the fibers for computation [12, 13].

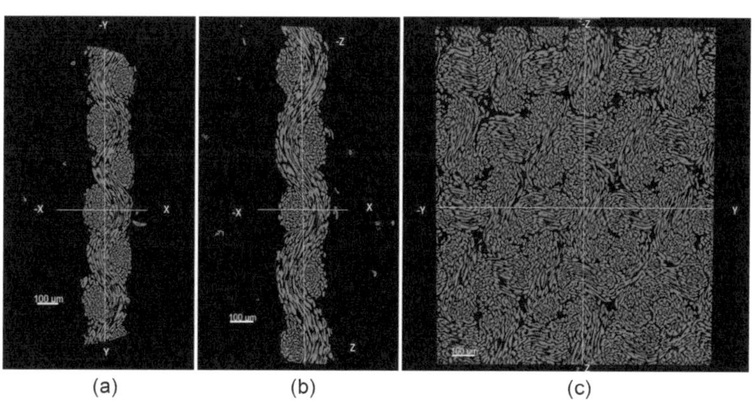

Figure 5.
2D views of P40H fabric where each dimension is perpendicular to each other. X, Y and Z-axis indicated to the thickness, warp and weft directions respectively [12, 13].

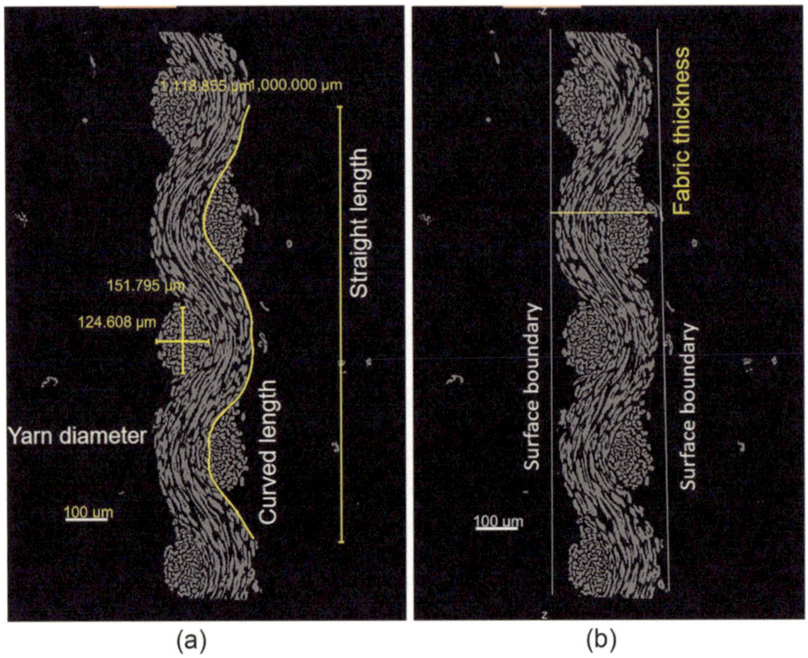

Figure 6.
Measurements of (a) yarn diameter and crimp% (requires the straight length and curved length), and (b) fabric thickness defining the two surfaces [12, 13].

Yarn Crimp: Yarn crimp (%) was calculated using the Eq. (9) below. Ten measurements of both straight length and curved length (as depicted in **Figure 6a**) were taken to calculate the respective warp crimp% (from XY images) and weft crimp% (from XZ images) [12, 13].

$$\text{Yarn crimp\%} = \frac{\text{Curved length} - \text{Straight length}}{\text{Straight length}} \times 100 \tag{9}$$

Yarn Packing Factor: Yarn packing factor was calculated using the yarn diameters obtained from the CT measurements (units converted as required) in Eq. (2) as discussed in Section 2.1 [12, 13].

Fabric Thickness: CT fabric thickness (µm) was defined and measured by the distance between the two surfaces (technical face and technical back) as depicted in **Figure 6b**. A total of fifty measurements were taken from different cross-sections of each fabric. The values were then averaged and converted to mm [12, 13].

Fabric Cover Factor: CT fabric cover factor was calculated using the CT yarn diameters for the respective fabrics, and following the Eqs. (6)–(8) [12, 13].

3. Comparison of yarn and fabric properties (non-destructive versus theoretical and classical standards)

This section will provide an understanding of the cotton-made yarn and fabric properties and how realistic the nondestructive measurements from the XRMCT images are.

The yarn and fabric characteristics obtained from the classical test standards are displayed in **Table 2**. In addition, **Figures 7** and **8** depict the comparison between theoretical calculation and experimental measurements obtained from CT images and classical test standards.

Yarn Diameter: Results showed that both the theoretical warp and weft yarn diameters [calculated from Eq. (1)] were statistically different (P-value <0.05) than the respective CT yarn diameters (**Figure 7a**). Irrespective of weave design, the calculated theoretical diameters were constant for the same count yarn as well as for both warp and weft yarns. Experimental CT measurements showed that the same count yarns resulted in different diameters of the warp and weft yarns in both plain and satin woven fabrics. Diameter in 40Ne yarns was higher than 60Ne yarns. In addition, given the same yarn count, higher yarn diameters were observed in satin fabrics than the plain-woven fabrics [12, 13].

Different weave designs and thread densities attributed to the distortion of the yarns as they interlaced with each other during weaving. Such distortion resulted in the differences in the warp and weft yarn diameters which can be understood from **Figure 9**.

Plain woven fabrics contained high number of interlacements. In contrast, satin fabrics had more floats or weft yarns which allowed more freedom to the yarns to move around and resulted in larger yarn diameters (**Figure 9**). The authors suggested that the CT images were capable of providing more realistic yarn diameters in the fabrics than the theoretical values [12, 13].

Yarn Packing Factor: **Figure 7b** shows that the different diameter yarns obtained from the CT measurements resulted in a variation in the yarn packing factors while the theoretical yarn packing factor was constant (0.6 for ring-spun yarns). Given the same yarn count, an increase in yarn diameter decreases the yarn packing factor which was not reflected from the theoretical values but from the CT measurements [12, 13].

Yarn Crimp: Yarn crimp% obtained from ASTM D3883 (option A) test method (**Table 2**) and CT method were compared (**Figure 7c**). For both warp and weft yarns, results from a two-sample t-test at 95% confidence interval showed that the measurements from the two methods were not statistically different (P-value of 0.43 and a higher R^2). Given the same yarn count and weave design, a further analysis of the CT data showed that warp crimp% was significantly higher than the weft crimp% (P-value <0.05). In addition, yarns in the plain-woven fabrics exhibited higher crimp% than the satin fabrics. The ASTM method of yarn crimp measurement involved extraction of yarns while the CT method was nondestructive. Therefore, the CT method can be reliably used to measure yarn crimp without compromising the structure [12, 13].

Fabric Thickness: Fabric thicknesses obtained from ASTM D1777–96(2002) method and CT method were compared. Results depicted in **Table 2** and **Figure 8a** suggested that the measurements from the two methods were not statistically different (P-value 0.21, and $R^2 = 0.97$). It was also observed that the satin fabrics exhibited higher thickness than the plain-woven fabrics. In the satin fabrics, the presence of more floats or weft yarns which extended across multiple yarns (**Figure 9**) attributed to overriding of yarns and looser fabric construction, and therefore, greater fabric volume and thickness than the plain-woven fabrics. In addition, 60Ne yarn-made satin fabrics had lower thickness than the 40Ne yarn-made fabrics. The smaller diameter 60Ne yarns contained higher twists (TPI) which compressed the fibers and increased yarn tightness (**Table 2**), and therefore, attributed to lower fabric thickness.

Fabric code	Yarn twist (TPI ± Std. Dev.)	Twist multiplier, $TM = \frac{TPI}{\sqrt{Ne}}$	Yarn crimp% ± Std. Dev.		Fabric thread density		Basis weight (gsm) ± Std. Dev.	Fabric thickness (mm) ± Std. Dev.	Calculated volumetric density (g/cm^3)
			Warp crimp %	Weft crimp %	EPI	PPI			
P40L	28.15 ± 1.57	4.45 ± 0.25	11.00 ± 0.03	10.50 ± 0.03	151	84	128.63 ± 0.64	0.26 ± 0.00	0.49
P40H	26.64 ± 2.08	4.21 ± 0.33	13.13 ± 0.03	11.88 ± 0.05	141	116	140.81 ± 1.34	0.24 ± 0.00	0.59
S40L	28.10 ± 0.84	4.44 ± 0.13	13.00 ± 0.04	6.88 ± 0.03	144	78	129.16 ± 0.46	0.42 ± 0.00	0.31
S40H	25.58 ± 0.81	4.04 ± 0.13	7.75 ± 0.04	4.00 ± 0.03	187	74	147.88 ± 0.30	0.34 ± 0.00	0.43
P60L	35.22 ± 2.03	4.55 ± 0.23	15.25 ± 0.05	7.50 ± 0.05	156	100	99.01 ± 1.75	0.2 ± 0.00	0.50
P60H	35.89 ± 0.98	4.63 ± 0.13	4.00 ± 0.03	15.25 ± 0.05	185	144	117.79 ± 0.95	0.28 ± 0.00	0.42
S60L	29.16 ± 1.92	3.76 ± 0.25	5.88 ± 0.03	3.00 ± 0.03	220	96	109.64 ± 0.74	0.24 ± 0.00	0.46
S60H	36.51 ± 2.10	4.71 ± 0.27	13.00 ± 0.03	4.75 ± 0.03	219	132	128.47 ± 0.50	0.28 ± 0.00	0.46

Table 2.
Structural characteristics of the yarns and fabrics used in the study [12, 13].

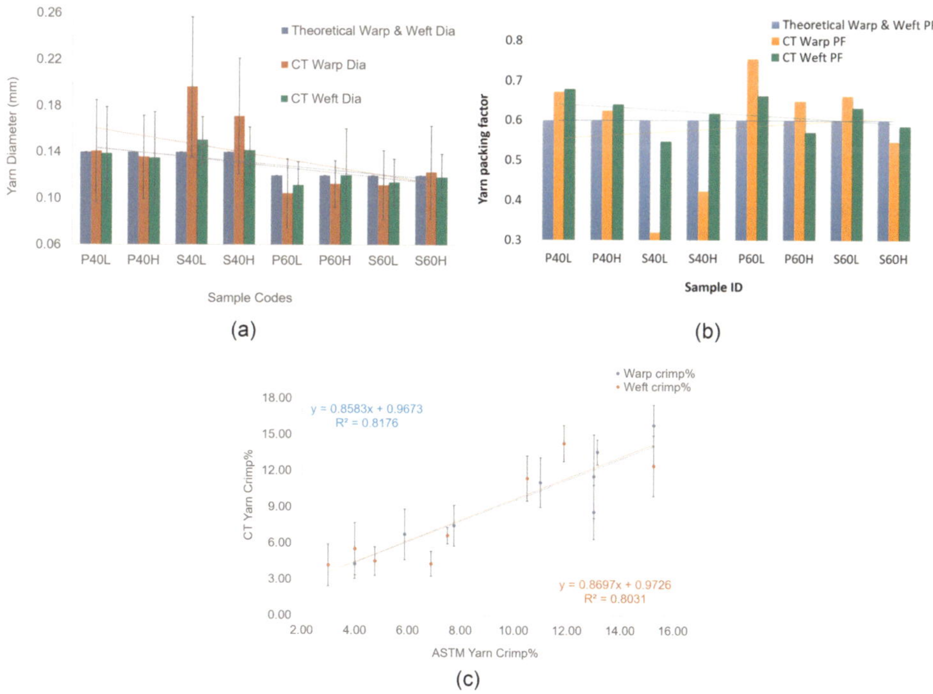

Figure 7.
Yarn characteristics obtained from ASTM test standards and XRM-CT methods. (a) Yarn diameter, (b) yarn packing factor (PF), and (c) yarn crimp% [12, 13].

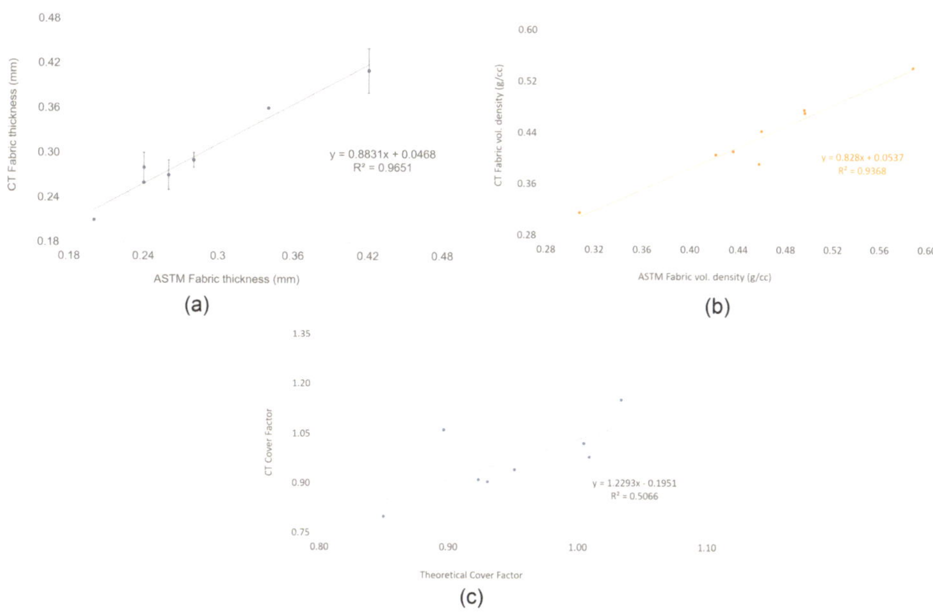

Figure 8.
Fabric characteristics obtained from ASTM test standards, theoretical values and XRM-CT method. (a) Fabric thickness, (b) fabric volumetric density calculated using ASTM thickness and CT thickness, and (c) fabric cover factor calculated using theoretical yarn diameters and CT yarn diameters [12, 13].

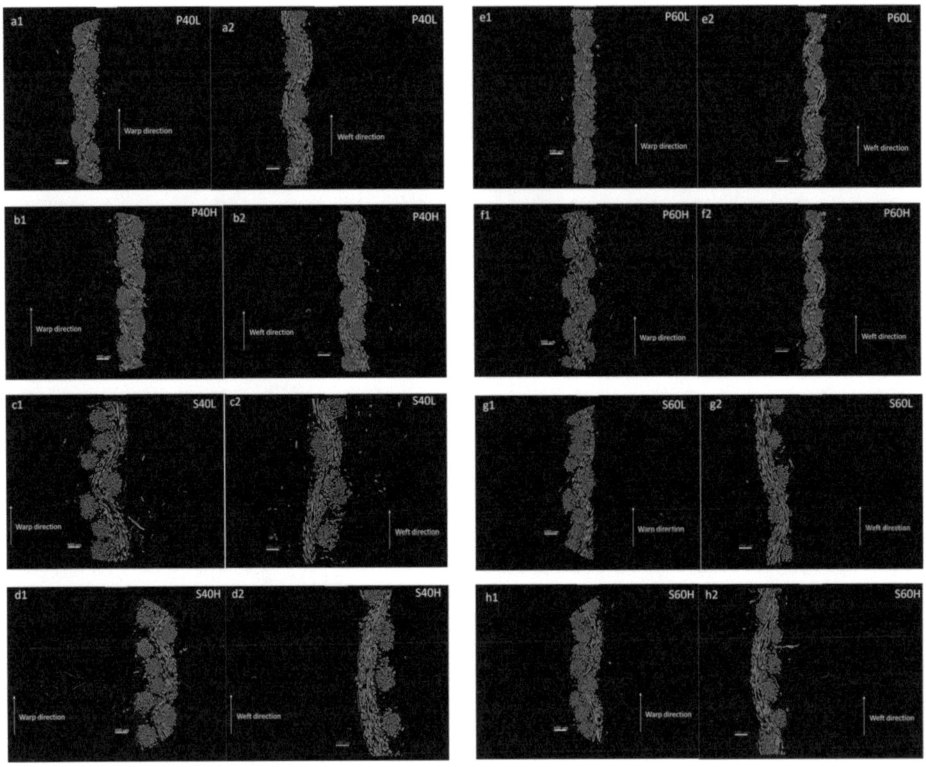

Figure 9.
X-ray micro-computed tomographic 2D images of the fabrics studied. (a) P40L, (b) P40H, (c) S40L, (d) S40H, (e) P60L, (f) P60H, (g) S60L, and (h) S60H [12, 13].

However, despite the different twists in different count-yarns in the fabrics, twist multiplier were nearly the same suggesting the same internal structure (helix angle) of the yarns [12, 13].

With increasing fabric density (thread density), thickness decreased in 40Ne yarn-made fabrics, and increased in 60Ne yarn-made fabrics (**Figure 9**). With increasing fabric density, less twisted and larger diameter 40Ne yarns compressed within the structure, and filled up the voids. This resulted in a closer fiber and yarn packing, and reduced thickness in the 40Ne yarn-made fabrics (**Figures 7b** and **9**). In contrast, finer, tighter (with more twists) and smaller diameter 60Ne yarns had less voids (as discussed above in yarn packing factor). Hence, in the presence of low voids in the 60Ne yarns, the yarns expanded in the thickness direction with increasing fabric density (**Figures 7b** and **9**) [12, 13].

Fabric Volumetric Density: Both the ASTM thickness and CT thickness were used to calculate fabric volumetric density (g/cm^3) for further comparison (**Figure 8b**), and were obtained by multiplying fabric basis weight (gsm) with thickness. The methods did not exhibit any statistical differences in the measurements (P-value 0.51) [12, 13].

Fabric Cover Factor: Fabric cover factor obtained from the theoretical values and the CT method (using CT yarn diameter) were compared and depicted in **Figure 8c**. A further statistical analysis showed that there was no significant difference between the measurements obtained from the two methods (*P*-value 0.62), and satin fabrics exhibited higher fabric cover factor. Irrespective of weave design, no significant difference was observed between 40 and 60Ne yarn-made fabrics [12, 13].

Therefore, unlike other classical test methods, the CT method are nondestructive and capable of providing realistic measurements from both the 2D and 3D images. Apart from the structural properties discussed in this Chapter, the CT method can be used for a more advanced analysis of the fabrics such as fiber area distribution [12, 13], porosity analysis [22], surface profile analysis [1] etc. Such in-depth analysis and understanding of the cotton-made yarn and fabric structures will help engineers and scientists optimize the structure and improve the performance of the textiles for diverse applications and end-uses.

4. Conclusion

In this chapter, a nondestructive characterization method using an X-ray microscope was discussed. The method involves acquiring a high-resolution tomographic 3D images of the cotton fabrics, and is capable of providing reliable quantitative measurements. Yarn diameter, packing factor, yarn crimp, fabric thickness, volumetric density and fabric cover factor were measured and discussed. The measurements were also compared with those obtained from existing theory and classical test methods. The findings on cotton-based yarns and fabrics discussed in this chapter will also assist in non-destructive characterizations and performance evaluations of other natural and synthetic fibers, yarns and fabrics as well as other material classes.

Acknowledgements

We sincerely appreciate the supports from the Analytical Instrumentation Facility (AIF) and the Wilson College of Textiles, North Carolina State University, USA. We acknowledge the technical advice from Anton Jansson (Ex-Postdoctoral Research Scholar, AIF). This work was performed in part at the Analytical Instrumentation Facility (AIF) at North Carolina State University, which is supported by the State of North Carolina and the National Science Foundation (award number ECCS-1542015). The AIF is a member of the North Carolina Research Triangle Nanotechnology Network (RTNN), a site in the National Nanotechnology Coordinated Infrastructure (NNCI).

Conflict of interest

The authors declare no conflict of interest.

Author details

Ruksana Baby[1], Kavita Mathur[2*] and Emiel DenHartog[2]

1 Analytical Instrumentation Facility, North Carolina State University, Raleigh, North Carolina, USA

2 Wilson College of Textiles, North Carolina State University, Raleigh, North Carolina, USA

*Address all correspondence to: kmathur@ncsu.edu

References

[1] Derler S, Rotaru G, Ke W, Issawi-Frischknecht LE, Kellenberger P, Scheel-Sailer A, Rossi RM. Microscopic contact area and friction between medical textiles and skin. Journal of the Mechanical Behavior of Biomedical Materials, 2014; **38**:114–125. DOI: 10.1016/j.jmbbm.2014.06.014

[2] Torbacke M, Rudolphi ÅK, Kassfeldt E. Skin Against Different Fabrics for Medical Use. 2012. Retrieved from: http://urn.kb.se/resolve?urn=urn:nbn:se:ltu:diva-16461

[3] Tasron D, Maiti R, Hemming M, Lewis R, Carré M. Frictional interaction between running sock fabrics and plantar aspect of first metatarsal head in different moisture conditions. Procedia Engineering. 2016;**147**:753-758. DOI: 10.1016/j.proeng.2016.06.328

[4] Tasron DN, Thurston TJ, Carré MJ. Frictional behaviour of running sock textiles against plantar skin. Procedia Engineering. 2015;**112**: 110-115. DOI: 10.1016/j.proeng.2015.07.184

[5] Dong Y, Kong J, Mu C, Zhao C, Thomas NL, Lu X. Materials design towards sport textiles with low-friction and moisture-wicking dual functions. Materials & Design. 2015;**88**:82-87. DOI: 10.1016/j.matdes.2015.08.107

[6] Carré M, Tasron D, Lewis R, Hashmi F. Investigating foot-sock friction: A comparison of two different methodologies. Procedia Engineering. 2016;**147**:759-764. DOI: 10.1016/j.proeng.2016.06.324

[7] Baussan E, Bueno M, Rossi R, Derler S. Analysis of current running sock structures with regard to blister prevention. Textile Research Journal. 2013;**83**(8):836-848. DOI: 10.1177/0040517512461698

[8] Rotaru G, Pille D, Lehmeier FK, Stämpfli R, Scheel-Sailer A, Rossi RM, et al. friction between human skin and medical textiles for decubitus prevention. Tribology International, 2013; **65**:91–96. DOI: 10.1016/j.triboint.2013.02.005

[9] Basal G, Ilgaz S. A functional fabric for pressure ulcer prevention. Textile Research Journal. 2009;**79**(16):1415-1426. DOI: 10.1177/0040517509105600

[10] Derler S, Rao A, Ballistreri P, Huber R, Scheel-Sailer A, Rossi RM. Medical textiles with low friction for decubitus prevention. Tribology International. 1974;**46**(1):208-214. DOI: 10.1016/j.triboint.2011.03.011

[11] Schwartz D, Magen YK, Levy A, Gefen A. Effects of humidity on skin friction against medical textiles As related to prevention of pressure injuries. International Wound Journal. 2018;**15**(6):866-874. DOI: 10.1111/iwj.12937

[12] Baby R, Mathur K, DenHartog E. Nondestructive quantitative evaluation of yarns and fabrics and determination of contact area of fabrics using the X-ray microcomputed tomography system for skin-textile friction analysis. ACS Applied Materials & Interfaces. 2021;**13**(3):4652-4664. DOI: 10.1021/acsami.0c18300

[13] Baby R. Micro- and macro-scale analysis of frictional interaction between human skin and woven fabrics [Ph.D. Dissertation]. North Carolina: North Carolina State University; 2021

[14] ASTM D3883 - 04. Standard Test Method for Yarn Crimp and Yarn Take

Up in Woven Fabrics. (n.d.). Retrieved November 23, 2021, from http://www.astm.org/Standards/D3883.html

[15] Soltani P, Johari MS, Zarrebini M. Effect of 3D fiber orientation on permeability of realistic fibrous porous networks. Powder Technology. 2014;**254**: 44-56. DOI: 10.1016/j.powtec.2014.01.001

[16] Chiesura G, Luyckx G, Voet E, Lammens N, Van Paepegem W, Degrieck J, et al. A micro-computed tomography technique to study the quality of fibre optics embedded in composite materials. Sensors. 2015;**15**(5): 10852-10871

[17] Zeng X, Mooney SJ, Sturrock CJ. Assessing the effect of fibre extraction processes on the strength of flax fibre reinforcement. Composites Part A: Applied Science and Manufacturing. 2015;**70**:1-7. DOI: 10.1016/j.compositesa.2014.12.004

[18] Toda M, Grabowska KE. Computed microtomography in the analysis of fiber migration in yarn. Autex Research Journal. 2013;**13**(1):28-32. DOI: 10.2478/v10304-012-0020-x

[19] Badel P, Vidal-Sallé E, Maire E, Boisse P. Simulation and tomography analysis of textile composite reinforcement deformation at the mesoscopic scale. Composites Science and Technology. 2008;**68**(12):2433-2440. DOI: 10.1016/j.compscitech.2008.04.038

[20] Shinohara T, Takayama J, Ohyama S, Kobayashi A. Automatic weave diagram construction from yarn positional data of woven fabric. Textile Research Journal. 2008;**78**(9):745-751. DOI: 10.1177/0040517507083525

[21] Shinohara T, Takayama J, Ohyama S, Kobayashi A. Extraction of yarn positional information from a three-dimensional CT image of textile fabric using yarn tracing with a filament model for structure analysis. Textile Research Journal. 2010;**80**(7):623-630. DOI: 10.1177/0040517509342320

[22] Zhu L, Wang X, Hinestroza JP, Naebe M. Determination of the porosity in a bifacial fabric using micro-computed tomography and three-dimensional reconstruction. Textile Research Journal. 2018;**88**(11):1263-1277. DOI: 10.1177/0040517517698987

[23] Laourine E, Cherif C. Characterization of barrier properties of woven fabrics for surgical protective textiles. AUTEX Research Journal. 2011; **11**(2):31-36

[24] Zankel A, Wagner J, Poelt P. Serial sectioning methods for 3D investigations in materials science. Micron. 2014;**62**:66-78. DOI: 10.1016/j.micron.2014.03.002

[25] Reingruber H, Zankel A, Mayrhofer C, Poelt P. Quantitative characterization of microfiltration membranes by 3D reconstruction. Journal of Membrane Science. 2011;**372**(1):66-74. DOI: 10.1016/j.memsci.2011.01.037

[26] Mathur K. Color prediction model for jacquard tapestry woven fabrics [Ph. D. Dissertation]. North Carolina: North Carolina State University; 2008

[27] Baby R. Wicking of small drops into knit fabrics [MS Thesis]. North Carolina: North Carolina State University; 2017

[28] Baby R, Michilesen S, Wu J. Effects of yarn size and blood drop size on wicking and bloodstains in textiles. Journal of Forensic Sciences. 2021;**66**(4): 1246-1256

[29] ASTM D1422-99(2008). Standard Test Method for Twist in Single Spun

Yarns by the Untwist-Retwist Method. (n.d.). Retrieved November 23, 2021, from https://www.astm.org/d1422-99r08.html

[30] ASTM 3775. (n.d.). Standard Test Method for End (Warp) and Pick (Filling) Count of Woven Fabrics. (n.d.). Retrieved November 23, 2021, from https://www.astm.org/d3775-17e01.html

[31] ASTM D3776. Standard Test Methods for Mass Per Unit Area (Weight) of Fabric. (n.d.). Retrieved November 23, 2021, from https://www.astm.org/d3776_d3776m-20.html

[32] ASTM D1777-96(2002). Standard Test Method for Thickness of Textile Materials. (n.d.). Retrieved November 23, 2021, from https://www.astm.org/d1777-96r02.html

Chapter 6

Absorbency and Wicking Behaviour of Natural Fibre-Based Yarn and Fabric

Palash Mallick and Susanta Sekhar De

Abstract

In textiles of natural fibres, absorbency and wicking are considered as most indigenous liquid transmission properties, which play significant role in assessing the comfort of textile garments as well as in functioning of technical textiles. To explain the general principles of absorbency and wicking and their related parameters, a review of different theories is illustrated. Various structural aspects of textile intermediates such as yarn and fabric on wicking are discussed. An experimental analysis is also described to understand the wicking phenomena on fabrics in multiple directions such as warp-way, weft-way and diagonally for varying weave forms. A critical evaluation is presented for establish the wicking behaviour between these fabrics and their constituted yarns having preferred twisted forms such as single and ply. A unique representation of correlation matrices is shown to interrelate the yarn and fabric wicking.

Keywords: natural fibre, absorbency, wicking, capillary action, fabric structure

1. Introduction

In terms of comfort and exuberant use, cotton fibre is considered the most classical fibre in garment technology. Versatility of cotton fibres ranging from very finer to very courser and structure of cotton yarns such as single to ply are considered as decisive factors of selection regarding their use in apparel production. On the other hand, the comfort property, being one of the prime features of woven fabric used as a garment, is greatly influenced by the structure of constituent yarns in that particular fabric. Comfort evaluation includes mainly vapour and liquid transmission properties of yarn and fabric at constant atmospheric conditions and constraints. Liquid transmission includes water absorbency and wicking [1, 2].

Absorbency is used to describe the ability of a fabric to take in moisture—an important feature that influences a variety of other factors such as skin comfort, static build-up, shrinkage, water repellency and wrinkle recovery. One of the most common occurrences in the manufacturing and usage of textile materials is liquid flow. Wicking is the spontaneous transfer of a liquid into a porous structure caused by capillary forces. Therefore, study of capillary flow in textile media is very important. When the free energy of the solid-gas interface surpasses that of the solid-liquid interface, capillary phenomena occur. A liquid that does not wet fibres cannot wick into a fabric, and wicking can only take place when fibres are assembled with capillary spaces. But the objective of this is to present an overview only in wicking property of textiles.

Study of wicking in textile material is of great interest for two main reasons. Firstly, it allows a better understanding of liquid-fibre contact in order to characterise any liquid flow of spin finishes, dyeing or coating of either fabrics or yarns. Secondly, it enables the characterisation of textile structures, their porosity resulting from the capillaries formed by the inter-filament spaces in which the liquid flows [3]. Moisture absorption is intimately linked to the comfort of garments made from cellulosic fibres. The term wetting is usually used to describe the displacement of solid-air interface with solid liquid interface. Wetting behaviour is commonly characterised by the value of the contact angle within the liquid [4]. The wetting and wicking behaviour of the fibrous structure is a critical aspect of performance of products such as sports clothes, hygiene disposable materials and medical products. Clothing comfort is influenced by the wetting and wicking processes that occur during wear.

Therefore, a thorough understanding of liquid transmission behaviour inside various intermediate forms of textiles is obviously needed through absorbency and wicking phenomena of yarns and fabrics. As the yarns are composed of fibres with a structured arrangement, any change of this arrangement by twisting and plying creates different sort of wicking behaviour on them and in turn, the fabrics made out of these yarns along with its varying weave forms and directions must exhibit different nature of wicking effects. The findings that are interlinking between fabric wicking and corresponding yarn wicking are always a matter of interest to study the nature of absorbency specially for textiles made out of natural fibres and hence, are elaborated progressively in this chapter.

2. Lucas-Washburn theory

The so-called wicking (or absorbency) rate is of great importance for both scientific and practical uses. The Lucas-Washburn idea provides a more scientific definition of the wicking rate. This theory deals with the rate of a liquid drawn into a circular tube via capillary action. A capillary like this is a severely simplified representation of a pore in a true fibrous media with a complex structure. For laminar viscous flows, theory is a specific form of the Hagen-Poiseuille law state Landau and Lifshitz. According to this law, the volume dV of a Newtonian liquid with viscosity μ that wets through a tube of radius r and length h during dt is given by the relation:

$$dV/dt = \pi r4\ (p1 - p2)/8h\mu \quad (1)$$

where $p1 - p2$ is the pressure difference between the tube ends. The capillary force and gravitation both contribute to the pressure difference here. The angle between the tube axis and the vertical direction is denoted by β, while the contact angle of the liquid against the tube wall is denoted by θ. The value of the capillary pressure p1 is:

$$p1 = 2\gamma\ \cos\ \theta/r \quad (2)$$

While hydrostatic pressure p2 is:

$$p2 = h\ \zeta\ g\ \cos\ \beta, \quad (3)$$

where γ denotes the liquid surface tension, ζ is liquid density, g is the gravitational acceleration, and h, in this case, is the distance travelled by the liquid measured from the reservoir along the tube axis. This distance obviously is the function of time, h = h (t), for a given system. The Lucas-Washburn equation is obtained by substituting the values p1, p2 and h(t) into Eq. (1), expressing the liquid volume in the capillary V as $\pi r^2 h$:

$$dV/dt = r\,\gamma\,\cos\,\theta/4\mu h – r2\zeta\,g\,\cos\,\beta/8\mu \tag{4}$$

For a given system, parameters such as r, γ, θ, ζ, g and β remain constant. We can then reduce the Lucas-Washburn equation (4) by introducing two constants,

$$K' = r\,\gamma\,\cos\,\theta/4\,\mu \text{ and } L' = r\,\zeta\,g\,\cos\,\beta/8\mu \tag{5}$$

into a simplified version,

$$dh/dt = K'/h – L' \tag{6}$$

The above relation is a non-linear ordinary differential equation that is solvable only after ignoring the parameter L'; this has a physical interpretation when either the liquid penetration is horizontal (β = 900 C) or r is small or the rising liquid height h is low that $K'/h >> L'$ or $L' \rightarrow 0$, and the effects of the gravitational field are negligible and the acceleration g vanishes. The Lucas-Washburn equation (6) could thus be solved with ease:

$$h = \sqrt{2K't}. \tag{7}$$

The result satisfies the initial condition h = 0 for t = 0.

Despite the intricate, noncircular, non-uniform and nonparallel nature of the pore spaces, the Lucas-Washburn technique provides an approximation tool for investigating the wicking and wetting behaviour of textiles. In the field of liquid sorption in a porous area, Washburn (1921) has expressed the following equation:

$$H = Ct^{0.5} \tag{8}$$

where H is the wicking height (m); C, the capillary liquid transport constant. The capillary force causes liquid to flow through a capillary channel, as shown in the equation above. The radius of the capillary channel, the contact angle between the liquid and the capillary channel and the rheological qualities of the liquid all influence the capillary force [5].

3. Wetting

The displacement of a solid-air interface with a solid-liquid contact is commonly referred to as 'wetting'. Two separate equilibrium regimes may be identified when a small liquid droplet comes into contact with a flat solid surface. Complete wetting with a zero-contact angle or partial wetness with a finite contact angle is shown in **Figure 1**. The equilibrium at a solid-liquid boundary is commonly described by the Young's equation:

$$\gamma_{SV} - \gamma_{SL} - \gamma_{LV} \cos\theta = 0 \tag{9}$$

where, γ_{SV}, γ_{SL} and γ_{LV} denote interfacial tensions between solid/vapour, solid/liquid and liquid/vapour, respectively, and θ is the equilibrium contact angle.

The parameter that distinguishes partial wetting and complete wetting is the so-called spreading parameter S, which measures the difference between the surface energy (per unit area) of the substrate when dry and wet:

$$S = [E_{substrate}]_{dry} - [E_{substrate}]_{wet} \tag{10}$$

Or,

$$S = \gamma_{SV} - (\gamma_{SL} + \gamma_{LV}) \tag{11}$$

The liquid spreads entirely to lower its surface energy ($\theta = 0$) if the parameter S is positive. The end result is a layer with a nano-scopic thickness that results from molecular and capillary forces competing. If S is negative, the drop does not spread out and instead forms an equilibrium spherical cap resting on the substrate with a contact angle of. When $\theta \leq \pi/2$, a liquid is said to be 'mainly wetting', while when $\theta > \pi/2$, it is said to be 'primarily non-wetting'. When a surface is connected with water, it is referred to as 'hydrophilic' when it is $\theta \leq \pi/2$ and 'hydrophobic' when it is $\theta > \pi/2$ [6].

3.1 Wicking

Wicking is the capillary-driven spontaneous flow of a liquid in a porous medium. Wetting causes capillary forces; hence, wicking is the result of spontaneous wetting in a capillary system. A meniscus is created in the simplest case of wicking in a single capillary tube as shown in **Figure 2**. A pressure difference across the curved liquid/vapour contact is caused by the surface tension of the liquid. The value for the pressure difference of a spherical surface was deduced in 1805 independently by Thomas Young and Pierre Simon de Laplace and is represented with the so-called Young-Laplace equation [6]:

$$\Delta P = \gamma_{LV}\left(\frac{1}{R1} - \frac{1}{R2}\right) \tag{12}$$

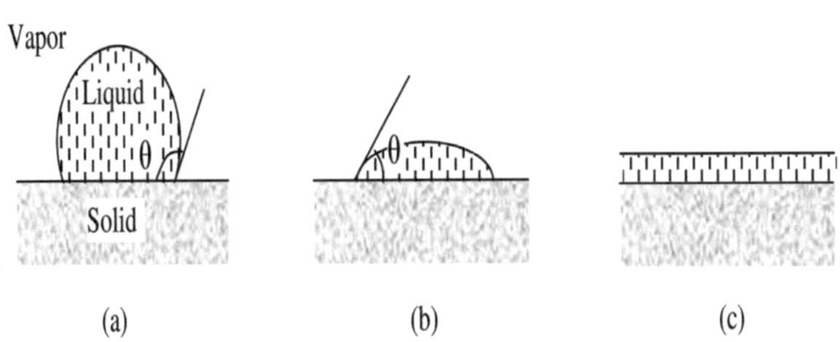

Figure 1.
A small liquid droplet in equilibrium over a horizontal surface: (a) partial wetting, mostly non-wetting; (b) partial wetting, mostly wetting; (c) complete wetting.

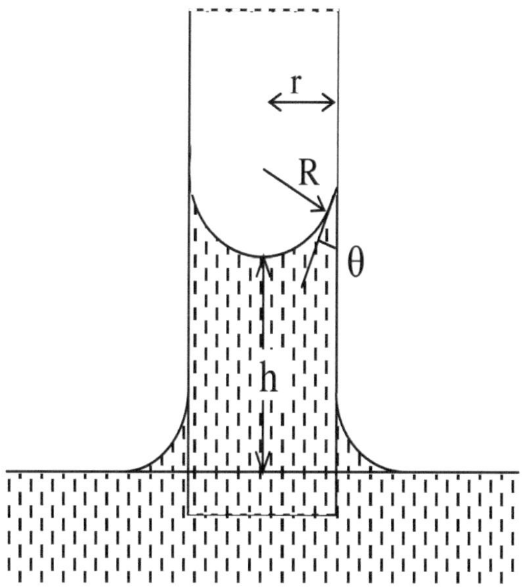

Figure 2.
Wicking in a capillary.

For a capillary with a circular cross section, the radii of the curved interface $R1$ and. $R2$ are equal. Thus:

$$\Delta P = 2\gamma_{LV}/R \tag{13}$$

where,

$$R = r/\cos\Theta \tag{14}$$

and r is the capillary radius. Since the capillary spaces in a fibrous assembly are not uniform, the effective capillary radius *re* is utilised instead, which is usually an indirectly determined parameter.

3.2 Capillary theory

Capillary action is governed by the properties of the liquid, the fibre surface wetting characteristics and the geometric configurations of the porous medium. In a capillary, liquid rises due to the net positive force (ΔP) across the liquid-solid interface.

$$\Delta P = P\text{–}\delta gh, \tag{15}$$

where δ = liquid density in g/cc, g = gravitational acceleration of 980.7 cm/s2, height of liquid rise in cm = h, the capillary pressure (p) is described by the initial wetting force (Fwi) in the capillary area (πri^2):

$$P = Fwi/\pi ri^2 = 2\,\pi\,ri\gamma\cos\theta/\pi ri^2 = 2\,\gamma\;\cos\;\theta \tag{16}$$

where, γ = liquid surface tension in dyne/cm.
ri = radius inside the capillary in cm and.
θ = liquid-solid contact angle.

Where capillary pressure (P) is greater than the weight of the liquid (δgh), the positive forces drive the liquid upward. Upon reaching equilibrium where P – δgh, the net driving force ΔP becomes zero. The liquid stops rising at the equilibrium height (h).

4. Liquid transport system

Water transport in yarns is only slightly influenced by the wetting properties of the individual fibre materials and depends mainly on the wetting behaviour of the whole yarn. The rate of water transport decreases as the roughness of the yarn increases due to the random arrangement of its fibres [7]. This is thought to be dependent on two factors that are directly related to capillary water transfer. (a) As the yarn roughness increases, so does the effective advancing contact angle of water on the yarn. (b) As the fibre arrangement gets more random, the capillary continuity generated by the yarn fibres appears to decrease. Most elements of water transport behaviour are accounted for by the penetration of capillaries created by the fibres in the yarns. Both the amount of water carried by the fabric and the distance that it travels in unit time are influenced considerably by the randomness of the arrangement of fibres in the yarns [8].

Perwuelz et al., on the other hand, worked on liquid organisation during capillary rise in yarns. It was found that capillary rise was dependent on the capillary diffusion coefficient of the yarn, and twist increase reduced the average value of the diffusion coefficients [9]. Twist retraction can cause fibres around the yarn centre to buckle when a high twist is applied into the yarn. This can damage the pore structures between fibres and influence the wicking behaviour [10]. Moreover, another study showed that wicking velocity increased with the increase in cross-sectional area of yarn and decrease in liquid viscosity [11]. In accordance with the literature, packing density of the yarns also influences wicking property. Compact yarns have higher packing density than conventional ring spun yarns, as a result of that they show lower wickability [12].

4.1 Wicking in yarn

Researcher investigated the wicking property of the yarns and found that the wicking behaviour of the yarns improved with the increase in the cross-sectional area due to a larger number of capillaries in yarns [13]. In the case of plied yarns, it is been found that the difference of wicking height of 2-ply and other higher plied yarns is more in finer yarn (40^s) as compared with coarser yarn (20^s) [14]. The difference of wicking height between 2-ply yarn and a higher number of yarns gradually decreases from finer yarn to coarser yarn [15].

4.1.1 Effect of yarn modification on water absorbency

1. It is evident that as the fibres become coarser, the water absorbency increases. The explanation for this could be due to coarser fibres' higher bending rigidity [16, 17] and a decrease in the number of fibres in the yarn cross section. As a result, the bulkiness of the yarn increases [18, 19]. Increased water absorbency is also linked to the availability of more pore volume in yarn structures [19].

2. It has been noticed that when yarn coarseness increases, cloth water absorbency increases at first, then declines. The explanation for this could be due to a

decrease in yarn structure compactness [20] and an increase in yarn bulkiness, which increases the availability of air spaces in the yarn structure, resulting in an increase in water absorbency [21, 22].

3. It has been discovered that when the sheath content of the yarn increases, fabric's water absorbency also increases. The explanation for this could be due to an increase in fabric openness [23] and yarn bulkiness when the sheath is removed, resulting in more air gaps in the yarn structure, which leads to an increase in water absorbency [20].

4.1.2 Effect of yarn modification on wicking characteristics

1. It is evident that as the fibre gets coarser, the fabric wicks more. The explanation for this could be due to coarser fibres' higher bending rigidity [18] and a decrease in the number of fibres in the yarn cross section. As a result, the bulkiness of the yarn increases [19, 23]. Increased wicking is caused by the availability of increased pore space within the yarn structure [24].

2. It has been noticed that when yarn coarseness increases, fabric wicking increases at first, then decreases. Because of the decrease in compactness of yarn structure [20] and rise in bulkiness of yarn, the size of the pores initially increases as the yarn becomes coarser, increasing the availability of air spaces in the yarn structure. Wicking increases as a result of this [25]. However, beyond a certain degree, the pore structure opens up, slowing the flow of water through the pores.

3. Fabric wicking is observed to increase when the proportion of sheath fibres rises. The explanation for this could be due to an increase in fabric openness [21] and yarn bulkiness when the sheath is removed, resulting in more air gaps in the yarn structure [23]. Furthermore, parallel fibre arrangement creates correct channelling within the yarn, resulting in increased wicking [26, 27].

4.2 Wicking in fabric

Fabric with a high and consistent absorbency is desirable in almost every wet finishing procedure and many finished fabrics. The wicking property of materials affects their absorbency. When a fabric is entirely or partially submerged in a liquid, or in touch with a small amount of liquid, such as a drop placed on the fabric, wicking occurs. As a result, capillary penetration of a liquid can occur from either an infinite (unlimited) or a finite (limited) reservoir. Immersion, transplanar wicking and longitudinal wicking are the wicking processes that occur from an infinite reservoir. A drop placed on the cloth surface demonstrates wicking from a restricted reservoir [28]. The places of interlacement between warps and wefts were constantly formed in horizontal and vertical directions as fluid flow split. The horizontal yarns (either warp or weft, depending on the test direction) served as fluid reservoirs to wick further along the vertical yarns [29].

4.2.1 Effect of fabric parameters on wicking and its application as technical textiles

Researchers have discovered that fabrics with equally distributed float wick slower and horizontally striped fabrics wick quicker. Because the floats of threads in horizontally striped fabrics are positioned on the edge of the horizontal stripes and spread throughout the entire fabric surface, this happens. This uneven structure could explain why horizontally striped materials have a higher wicking rate. The pace of warp and weft way wicking differs slightly, and in some situations, the rate of weft way wicking is larger than the rate of warp way wicking. This could be caused due to tension differences between warp and weft threads. The rate of wicking for uniformly dispersed floats increases as weave factor P1 increases, which is due to the increase in floats in the fabrics [30]. Fabrics made of ring yarn wick faster than fabrics made of compact yarn [31]. Moisture management behaviour of knitted fabric has been studied from structurally modified ring and vortex spun yarn where the air permeability, water vapour permeability and total absorbency of the knitted fabric made from changed yarn all improved significantly, whereas the wicking characteristic decreased [32]. Fabrics composed of coarser yarns (40s Ne) wick more quickly than fabrics made of finer yarns (50s Ne). Fabrics behave in the same way that yarns do when it comes to wicking. The difference in wickability between plain weave and twill weave fabrics grows as the constituent yarn fineness increases. This variation is substantially more noticeable in diagonal wicking than in warp- and weft-way wicking [33]. The Lucas-Washburn equation was found to be suitable for analysing the wicking behaviour of woven cotton fabrics, and the wicking height square had a positive and good correlation with time in both the warp and weft directions, according to Zhu et al. [34]. Wicking height decreases with increasing weft yarn density as it leads to a decrease in porosity due to increased warp yarn crimp [35].

Using a sink time technique, the wettability of various terry towel fabrics was tested, and it was discovered that fabrics with higher loop density (repeat with 3-pick) sink faster than those with lower pick density (4, 5 and 6-pick weaves). Vertical wicking studies reveal that materials with a 3-pick density wick the fastest, followed by fabrics with 4, 5 and 6-pick density [36]. Moisture management is used in applications such as active athletics, exercise garments, work clothing, intimate apparel and footwear to avoid or limit liquid collection on the wearer's skin owing to perspiration. This is accomplished by rapidly wicking or diffusing the liquid through an inner hydrophobic fibre layer to an outside hydrophilic layer, then evaporating it into the atmosphere. Sportwool is a two-layer sportswear with good moisture management qualities that is utilised in a variety of sports apparel. The inner layer, which is closest to the skin, is made up of chemically treated ultrafine Merino wool fibres ($<$20 μm). The exterior layer is completely constructed of polyester filament yarn. Wool fibres have a high-water vapour permeability, which allows heat and moisture to be transferred from the skin to the outside surface, where it can escape due to wind speed and body movement [37]. A multi-layer construction is appropriate for sportswear [38, 39]. According to D'Silva and Anand [38], in a two-layer construction, the wicking layer is made up of synthetic fibres, such as micro-denier polyester, while the absorbent and evaporating layer is commonly cotton or rayon.

Many healthcare and personal hygiene items require liquid flow management in terms of liquid volume and flow rate elements that are acceptable for the end use. The incorporation of effective capillary systems into the final product allows for controlled liquid flow. Hygiene goods are made up of different layers of materials with different properties that allow liquid or moisture to drain swiftly away from the inner surface of the material in contact with the wearer's body. Wicking and wetting principles are applied to multilayer protective clothing. The outer layer of protective clothing consists of permeable fabric or a liner or a layer made of carbon-loaded foam. To reduce the wetting due to body fluid, the outer layer is treated with a fluoro-chemical, and this reduces the surface energy, and thereby, the wetting is reduced [40]. Geotextiles are permeable fabrics that can hold solid items in place while allowing water to pass through. Due to their low cost, consistent qualities and convenience of placement, they have been widely used as drainage, separation and filter materials in geotechnical and geo-environmental activities for about 30 years. The primary goal of geotextiles is to allow water to flow through the filter into the drain over the course of the project while keeping soil particles in place and avoiding migration through the filter [40]. Natural geotextiles are now being used such as jute geotextiles (JGT) where biodegradability of JGT is an advantage when control of surficial soil erosion is considered. It acts as mulch, attenuates extremes of temperature, adds micronutrients to the soil, leaves fibrous residues that improve hydraulic conductivity of soil and thus eases dissipation of pore water pressure [41].

5. Experimental

The study focused on 100% cotton fabrics and their constituent cotton yarns, which included plied yarn for the warp and single yarn for the weft. During the preparation of the yarns, three (3) different counts, namely 20s, 30s and 40s Ne, were prepared from 0.9 s Ne cotton roving using a TRYTEX Miniature Ring Spinning machine for specific TM values of 4 (in indirect cotton count system), ensuring that the helix angle of fibres remained constant in all yarn structures. For all of the yarns, machine parameters such as the ring frame machine's spindle speed were set to 10,000 rpm.

Second, after necessary assembly winding and a portable laboratory winder, the yarns in the 20s, 30s and 40s were made 2-ply each using a TRYTEX Miniature TFO machine. Then, using 2-ply of each count as warp and single yarn of each count as weft, 1/1 plain and 2/1 twill weave fabric samples were produced, with their respective sample codes provided in **Table 1**.

Thus, a total of six yarn samples and six fabric samples were prepared for this study. All the yarn (single and plied) and fabric (plain and twill) samples had undergone a scouring process with the standard recipe as wicking is highly correlated to wetting so it is a prerequisite that the sample should be as hydrophilic as much as possible. All the scoured samples (yarns and fabrics) were subsequently used for testing of wicking phenomena after conditioning the fabric in standard atmospheric conditions for 48 h. The basic parameters of fabrics are determined in **Table 2**.

After setting up the experimental set-up in the testing centre, all of the yarn and fabric samples were evaluated for wicking parameters. Different yarn and fabric samples are initially built up in accordance with DIN 53924. Fabric pieces of 200 mm × 25 mm were used in the experiment. A beaker was placed beneath the yarn

Sample No.	Fabric construction	Sample code	
		Plain weave	Twill weave
1	2/20 as warp and 20s as weft	FP1	FT1
2	2/30 as warp and 30s as weft	FP2	FT2
3	2/40 as warp and 40s as weft	FP3	FT3

Table 1.
Fabric sample and used codes.

Sample No.	Sample codes	Thread density		Cloth cover factor
		Ends/inch	Picks/inch	
1	FP1	53	60	22.17
2	FP2	58	60	20.07
3	FP3	62	60	18.64
4	FT1	54	63	22.57
5	FT2	58	63	20.33
6	FT3	59	63	18.46

Table 2.
Thread density and cloth cover factor of respective fabric samples.

and fabric strip such that the yarn and fabric strip's end was dipped into the water. **Figures 3** and **4** demonstrate how yarns and fabric samples were hung in a stand by a clip. The laboratory travelling microscope was placed at the water level, and readings of the water level were taken at regular intervals.

6. Wicking behaviour of woven fabrics and its interrelation with their constituent yarns

As reviewed, wicking behaviour has been evaluated classically in terms of wicking height of the respective fabrics and its constituent yarns. Wicking height of fabrics has been shown in three different directions, and ultimately, related results have been used to establish the relationship between fabric wicking and its constituent yarn wicking phenomena.

6.1 Effect on wicking height (with respect to time) in various directions (warp-way, weft-way and diagonal-way) of fabrics as well as their constituent yarns

Tables 3–5 indicate the values of wicking height as a function of time as recorded in the trials for different orientations of fabrics and their constituent yarns in the 20s, 30s and 40s, respectively. **Figure 5** depicts the liquid flow for a specific set of warp and weft threads in the warp, weft and diagonal directions of fabric samples.

The tabulated findings show a few generalised observations about the behaviour of yarn wicking in comparison to fabric wicking. First and foremost, it is well established that ply yarns usually provide superior wicking than single yarns of the same fineness due to the presence of densely populated fibres in the plied structure versus a single form of yarn. It is also discovered that the wicking height of constituent yarns, whether plied (used as a warp) or single (used as a weft), of any fineness (expressed by counts), is higher than that of their respective fabrics, regardless of the direction of testing for plain weave samples. This usual phenomenon can be understood specifically for plain weave due to the compact fibrous structure of yarns in comparison to the porous-interlaced structure of fabrics made out of this constituent yarn resulting in much better capillary action facilitating higher liquid transmission inside the structure. An exception to this trend has been observed for twill fabrics produced from medium and finer count yarns when viewed with respect to wicking of single yarn

DOI: http://dx.doi.org/10.5772/intechopen.102584

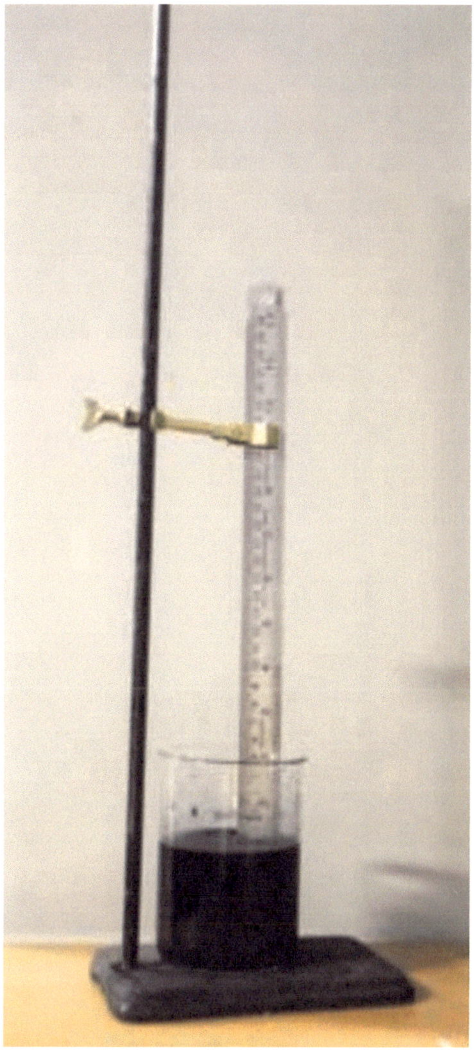

Figure 3.
Wicking set-up for yarn.

structure. Lesser number of interlacements may predominate for the attainment of higher wicking height in a twill weave in comparison to that of a single yarn.

However, a thorough explanation of the relative wicking phenomena is needed to set up a relationship between fabric wicking in various directions with respect to their constituent yarns of different counts. In this direction, a precise observation from **Table 3** reveals that 20s 2-ply yarn shows maximum wicking than single yarn of the same count when tested individually as well as than the fabrics made out of these yarns as warp and weft respectively irrespective of warp, weft and diagonal way-wicking test. In the case of 20s 2-ply yarn, wicking height reaches up to 13.7 cm, whereas in warp-way fabric, wicking height reaches only up to 4.9 cm, the single yarn is in the intermediate height of 8.2 cm. Fabric wicking is reduced in comparison to yarn wicking for three reasons: yarn deformation in the fabric due to crimped condition, the number of interlacement in a unit length of a fabric sample and finally, the

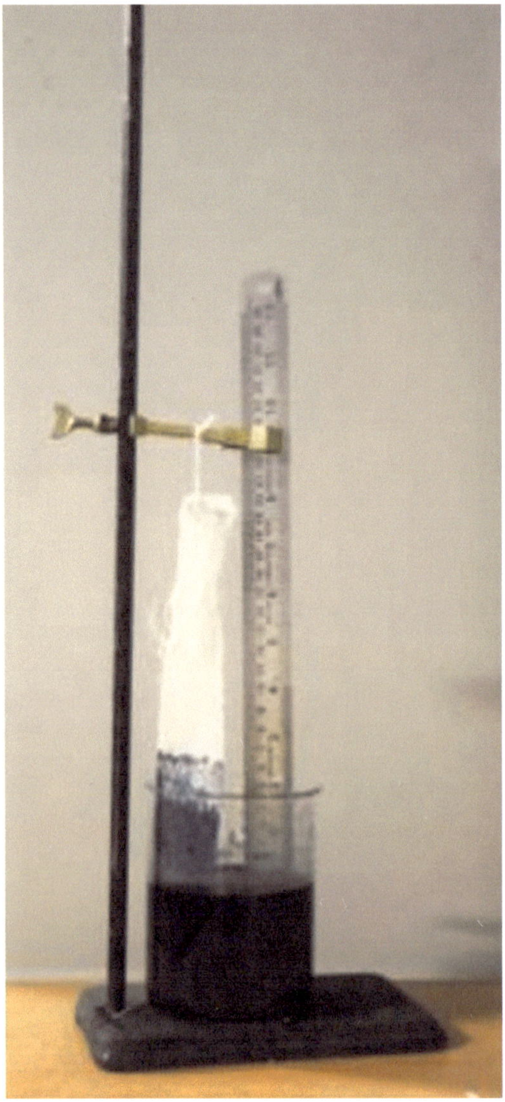

Figure 4.
Wicking set-up for fabric.

disruption of vertical wicking due to the occurrence of horizontal wicking at every point of interlacement. These characteristics can also be seen in twill weave fabrics. However, the effect is more or less the same as plain weave fabric considering the use of constituent coarser yarn in the fabric structure.

The maximum wicking height in 2 h for 30s 2-ply yarn is 10.7 cm, as indicated in **Table 4**, but when this yarn is present in fabric, the maximum warp-way wicking height is 5.7 cm in plain and 8.2 cm in a twill structure. The reason is the same as discussed before, but if we compare between plain weave fabric and twill weave fabric, the wicking height of twill weave fabric generates a higher value than that of the plain weave fabric. The explanation for this could be due to their fundamental design, as twill weave fabrics have less interlacements than plain weave fabrics. Another aspect we noticed from **Table 4** is that twill weave fabric's warp-way wicking reaches the height

Time (min)	Wicking height of yarn (cm)		Wicking height of FP1 (cm)			Wicking height of FT1 (cm)		
	2ply	Single	Warp	Weft	Diagonal	Warp	Weft	Diagonal
0.5	3.2	2.2	0.3	0.3	0.2	0.5	0.2	0.3
1	4.4	3.0	0.6	0.4	0.4	1.0	0.5	0.6
2	5.9	3.7	1.4	0.9	0.7	1.7	1.1	1.3
3	6.9	4.5	2.0	1.4	1.0	2.3	1.5	1.6
4	7.6	5.2	2.6	2.1	1.4	2.6	2.0	1.9
5	8.1	5.7	3.0	2.4	1.8	3.0	2.4	2.5
10	10.2	7.3	3.5	2.7	2.3	3.6	2.7	2.9
20	12.3	8.1	4.1	3.0	2.5	4.0	2.9	3.1
30	13.7	8.2	4.4	3.2	2.7	4.2	3.1	3.3
60	13.7	8.2	4.9	3.4	3.0	4.5	3.2	3.4
120	13.7	8.2	4.9	3.4	3.1	4.5	3.2	3.4

Table 3.
Wicking height of fabric samples (with codes) in different directions along with its constituent 2/20s Ne plied yarns (used as warp) and 20s Ne single yarns (used as weft).

Time (min)	Wicking height of yarn (cm)		Wicking height of FP1 (cm)			Wicking height of FT1 (cm)		
	2ply	Single	Warp	Weft	Diagonal	Warp	Weft	Diagonal
0.5	3.2	2.2	0.3	0.3	0.2	0.5	0.2	0.3
1	4.4	3.0	0.6	0.4	0.4	1.0	0.5	0.6
2	5.9	3.7	1.4	0.9	0.7	1.7	1.1	1.3
3	6.9	4.5	2.0	1.4	1.0	2.3	1.5	1.6
4	7.6	5.2	2.6	2.1	1.4	2.6	2.0	1.9
5	8.1	5.7	3.0	2.4	1.8	3.0	2.4	2.5
10	10.2	7.3	3.5	2.7	2.3	3.6	2.7	2.9
20	12.3	8.1	4.1	3.0	2.5	4.0	2.9	3.1
30	13.7	8.2	4.4	3.2	2.7	4.2	3.1	3.3
60	13.7	8.2	4.9	3.4	3.0	4.5	3.2	3.4
120	13.7	8.2	4.9	3.4	3.1	4.5	3.2	3.4

Table 4.
Wicking height of fabric samples (with codes) in different directions along with its constituent 2/30s Ne plied yarns (used as warp) and 30s Ne single yarns (used as weft).

of 30s single yarn, which is not visible in plain weave fabric. The comparatively finer yarn composition could be the predominant factor for this behaviour.

The wicking behaviour of plain weave fabric is low when compared with both ply yarn and single yarn, as shown in **Table 5**, but when considering the twill weave, the wicking height in diagonal-way direction crosses the wicking height of warp-way direction, crosses the wicking height of single yarn and approaches the wicking height

Time (min)	Wicking height of yarn (cm)		Wicking height of FP1 (cm)			Wicking height of FT1 (cm)		
	2ply	Single	Warp	Weft	Diagonal	Warp	Weft	Diagonal
0.5	3.2	2.2	0.3	0.3	0.2	0.5	0.2	0.3
1	4.4	3.0	0.6	0.4	0.4	1.0	0.5	0.6
2	5.9	3.7	1.4	0.9	0.7	1.7	1.1	1.3
3	6.9	4.5	2.0	1.4	1.0	2.3	1.5	1.6
4	7.6	5.2	2.6	2.1	1.4	2.6	2.0	1.9
5	8.1	5.7	3.0	2.4	1.8	3.0	2.4	2.5
10	10.2	7.3	3.5	2.7	2.3	3.6	2.7	2.9
20	12.3	8.1	4.1	3.0	2.5	4.0	2.9	3.1
30	13.7	8.2	4.4	3.2	2.7	4.2	3.1	3.3
60	13.7	8.2	4.9	3.4	3.0	4.5	3.2	3.4
120	13.7	8.2	4.9	3.4	3.1	4.5	3.2	3.4

Table 5.
Wicking height of fabric samples (with codes) in different directions along with its constituent 2/40s Ne plied yarns (used as warp) and 40s Ne single yarns (used as weft).

Figure 5.
Wicking phenomena in fabric for various directions such as (a) warp way, (b) weft way and (c) diagonal way.

of ply yarn. This phenomenon might be caused by the angle of ply yarn and single yarn orientation in diagonal-way fabrics. The yarns are at a 45° angle in this orientation, which facilitates the wicking process rather than creating obstructions in the point of interlacement as observed in warp or weft-way fabrics. Moreover, it is observed that the fabric made of 40s 2-ply warp and 40s weft twill weave fabric has higher wicking height, which is may be due to the reason of higher porosity of the fabric woven from the finer yarns. As the cover factor is low in the finer fabric, which may be responsible for higher porosity, which confirms the same reported by Erdumlu and Saricam [42]. According to the capillary principle, smaller pores are filled first and influence the movement of the liquid. As the smaller pores are completely filled, the liquid moves to the larger pores [43].

6.2 The relationship between yarn wicking and fabric wicking in reference to different constituent yarns for various directions of plain and twill fabrics

The accompanying graphical figures (**Figures 6–17**) show the relationship between fabric wicking in various directions (warp, weft and diagonal) and its constituent yarn

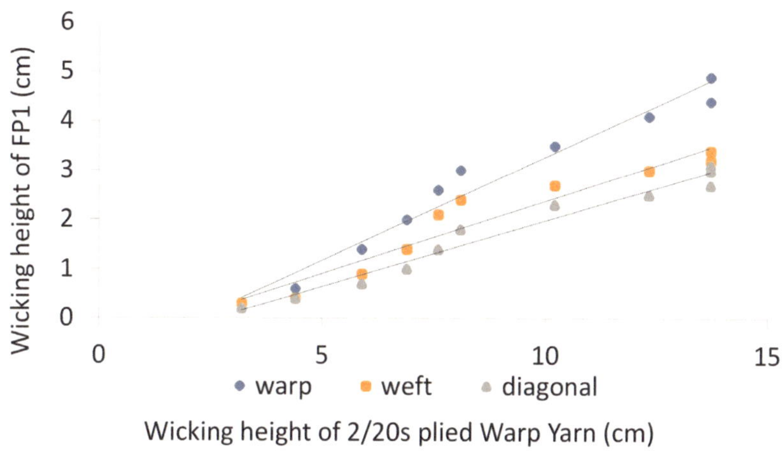

Figure 6.
Relationship of wicking height between plain weave fabric (FP1) and constituent plied warp yarn (2/20^s).

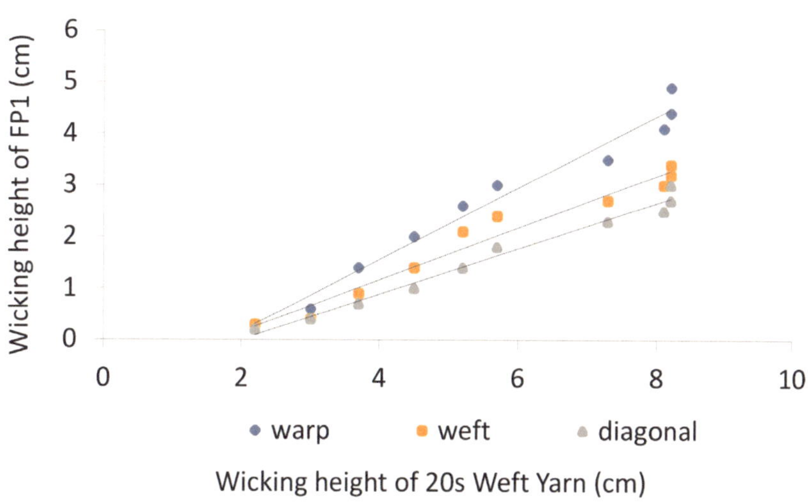

Figure 7.
Relationship of wicking height between plain weave fabric (FP1) and constituent single weft yarn (20^s).

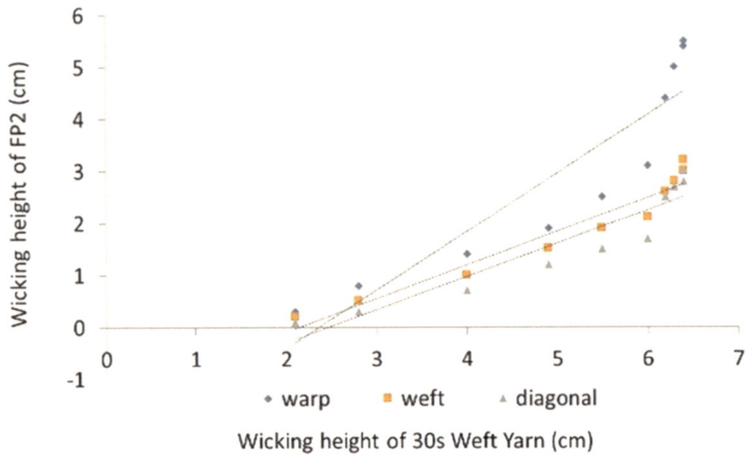

Figure 8.
Relationship of wicking height between plain weave fabric (FP2) and constituent plied warp yarn (2/30^s).

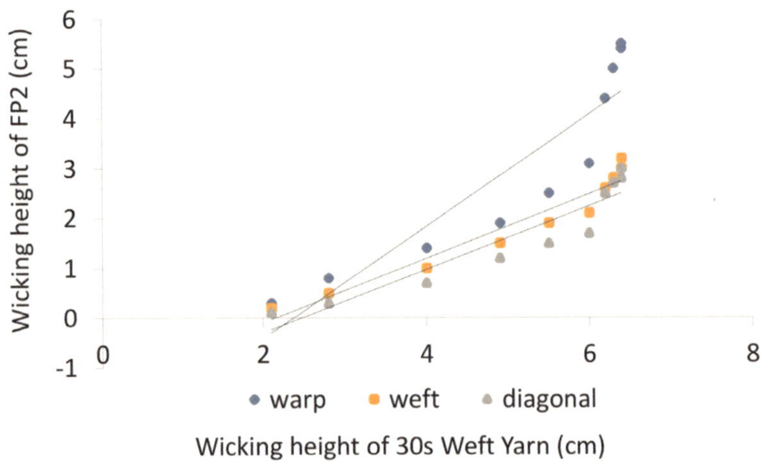

Figure 9.
Relationship of wicking height between plain weave fabric (FP2) and constituent single weft yarn (30^s).

wicking (warp or weft) for plain and twill weave structures, from coarser to finer count. The first six graphical figures (**Figures 6–11**) are for plain weave fabrics only, while the remaining (**Figures 12–17**) are for twill weave fabrics. For all of these representations, a linear relationship is found to be the best fit, and the associated regression equation is also derived to establish the relationship.

Table 6 has a relationship matrix containing all of the regression equations (18 numbers) from **Figures 4** to **9**, expressing substantial evidence of a highly positive relationship between fabric wicking (y) and yarn wicking (x) for plain weave samples. The coefficient of determination (*R*2) is found to be in the range of 0.836–0.991, which means the coefficient of correlation values are almost closer to 1, which indicates that the proportion of variation is very low in the dependent variable that can be attributed to the independent variable within the range of experiments. The R2 values

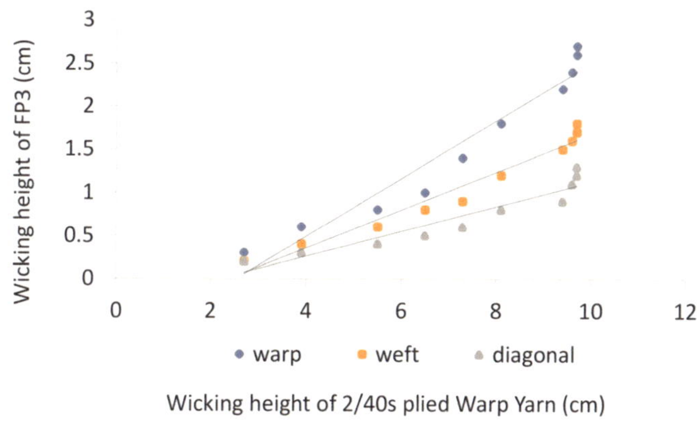

Figure 10.
Relationship of wicking height between plain weave fabric (FP3) and constituent plied warp yarn ($2/40^s$).

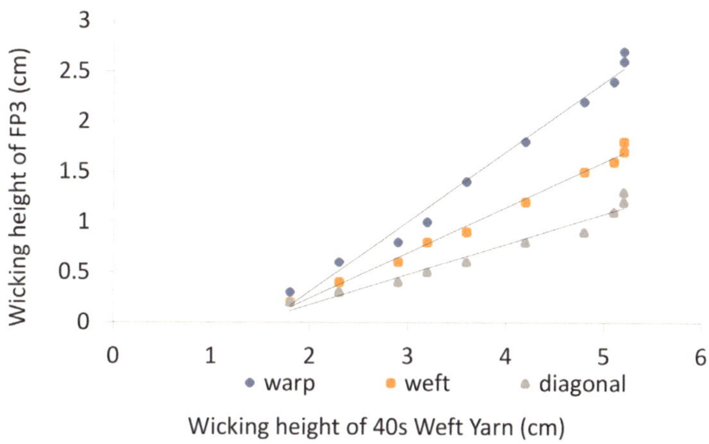

Figure 11.
Relationship of wicking height between plain weave fabric (FP3) and constituent single weft yarn (40^s).

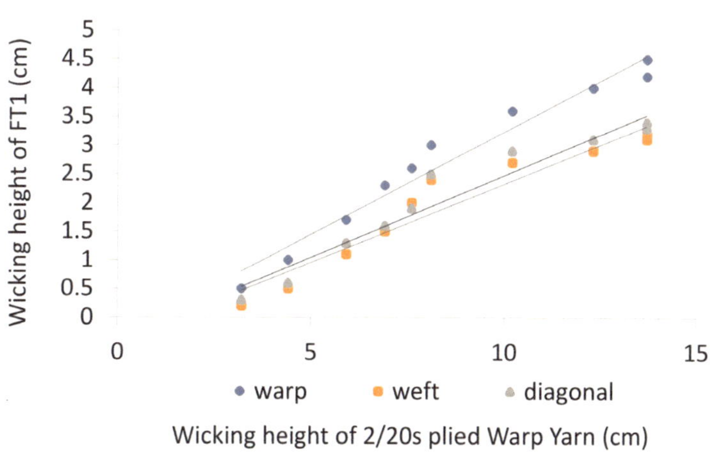

Figure 12.
Relationship of wicking height between twill weave fabric (FT1) and constituent plied warp yarn ($2/20^s$).

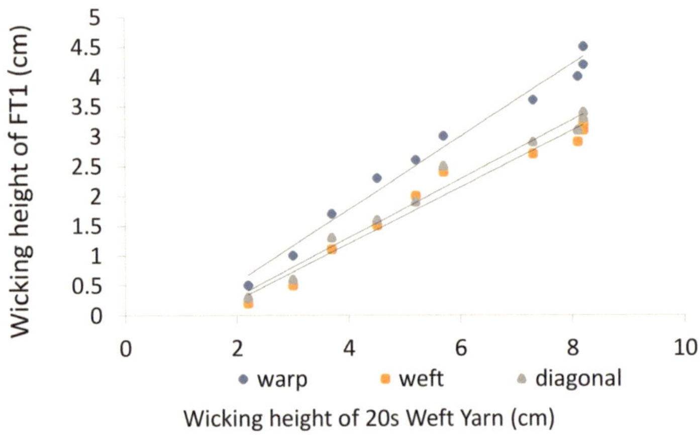

Figure 13.
Relationship of wicking height between twill weave fabric (FT1) and constituent single weft yarn (20^s).

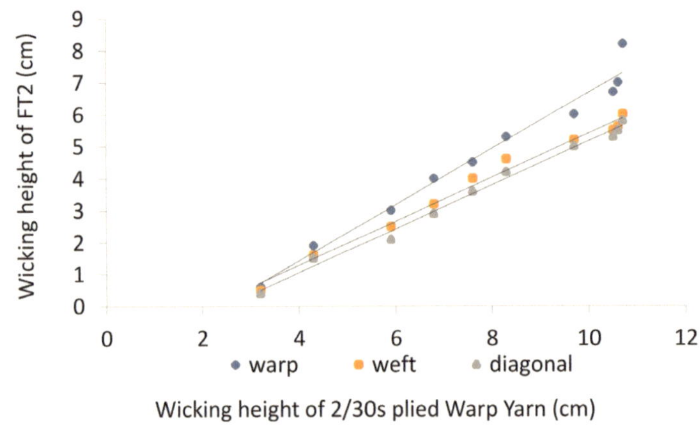

Figure 14.
Relationship of wicking height between twill weave fabric (FT2) and constituent plied warp yarn (2/30^s).

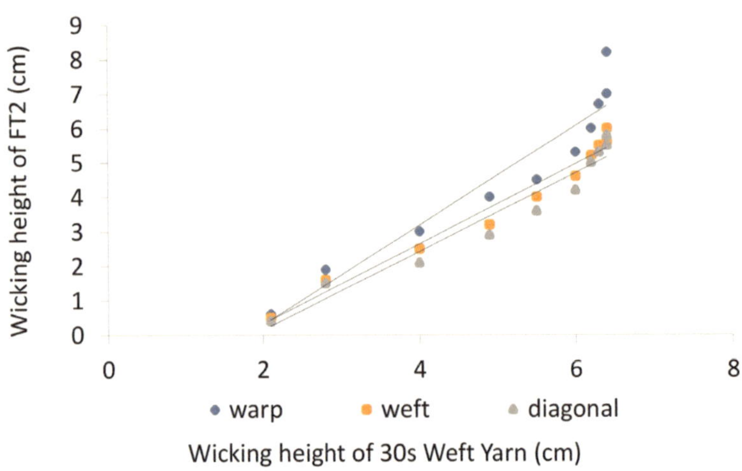

Figure 15.
Relationship of wicking height between twill weave fabric (FT2) and constituent single weft yarn (30^s).

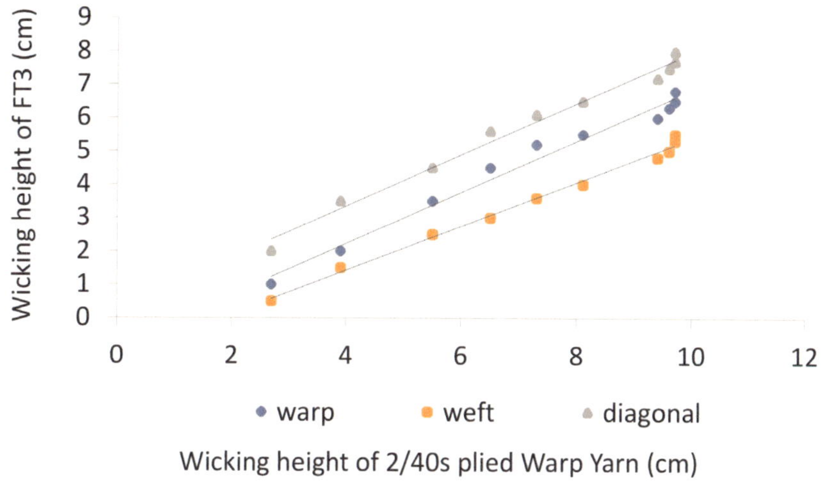

Figure 16.
Relationship of wicking height between twill weave fabric (FT3) and constituent plied warp yarn (2/40ˢ).

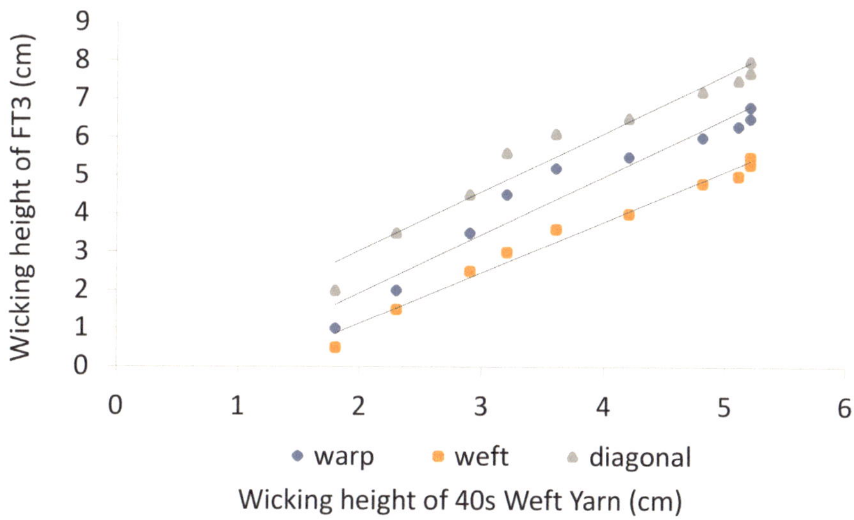

Figure 17.
Relationship of wicking height between twill weave fabric (FT3) and constituent single weft yarn (40ˢ).

of the diagonal-way samples are between the R2 values of warp-way and weft-way samples composed of coarser to medium fineness yarns (20s and 30s), most likely due to the resultant effect of both the constituent plied warp and single weft yarns because the yarns are not in the direction in which the wicking has been measured. Regardless of warp or weft yarns, the relationship in a diagonal direction is found to be inferior for finer count (40s) fabric samples. As seen in **Table 2**, this result might be attributable to the lower fabric cover.

The correlation of wicking behaviour between fabrics of different directions (warp-way, weft-way and diagonal way) with respect to their constituent warp and weft yarns for twill weave samples is shown in a single representation using a similar relationship matrix derived from **Figures 10–15** and arranged as **Table 7**. This table

Parameters		Wicking height of plain weave fabric (cm)		
		Warp-way	Weft-way	Diagonal-way
Wicking height of 20^s yarn (cm)	2-ply warp	y = 0.420x – 0.925 (R^2 = 0.972)	y = 0.294x – 0.562 (R^2 = 0.937)	y = 0.268x – 0.696 (R^2 = 0.969)
	Single weft	y = 0.713x – 1.289 (R^2 = 0.973)	y = 0.508x – 0.864 (R^2 = 0.968)	y = 0.457x – 0.938 (R^2 = 0.976)
Wicking height of 30^s yarn (cm)	2-ply warp	y = 0.709x – 2.478 (R^2 = 0.966)	y = 0.390x – 1.147 (R^2 = 0.990)	y = 0.393x – 1.403 (R^2 = 0.981)
	Single weft	y = 1.125x – 2.665 (R^2 = 0.836)	y = 0.645x – 1.384 (R^2 = 0.932)	y = 0.635x – 1.567 (R^2 = 0.881)
Wicking height of 40^s yarn (cm)	2-ply warp	y = 0.335x – 0.847 (R^2 = 0.941)	y = 0.219x – 0.520 (R^2 = 0.958)	y = 0.144x – 0.315 (R^2 = 0.895)
	Single weft	y = 0.695x – 1.082 (R^2 = 0.985)	y = 0.452x – 0.664 (R^2 = 0.991)	y = 0.335x – 0.847 (R^2 = 0.941)

Table 6.
Matrices of regression equations and R^2 values between fabric wicking and yarn wicking for plain weave fabric samples.

Parameters		Wicking height of twill fabric (cm)		
		Warp-way	Weft-way	Diagonal-way
Wicking height of 20^s yarn (cm)	2-ply warp	y = 0.356x – 0.335 (R^2 = 0.967)	y = 0.273x – 0.404 (R^2 = 0.931)	y = 0.285x – 0.376 (R^2 = 0.946)
	Single weft	y = 0.610x – 0.671 (R^2 = 0.982)	y = 0.473x – 0.696 (R^2 = 0.970)	y = 0.492x – 0.671 (R^2 = 0.979)
Wicking height of 30^s yarn (cm)	2-ply warp	y = 0.877x – 2.085 (R^2 = 0.974)	y = 0.687x – 1.466 (R^2 = 0.987)	y = 0.685x – 1.691 R^2 = 0.991
	Single weft	y = 1.446x – 2.600 (R^2 = 0.912)	y = 1.159x – 1.997 (R^2 = 0.966)	y = 1.140x – 2.139 (R^2 = 0.943)
Wicking height of 40^s yarn (cm)	2-ply warp	y = 0.769x – 0.84 (R^2 = 0.981)	y = 0.659x – 1.206 (R^2 = 0.991)	y = 0.771x + 0.272 (R^2 = 0.985)
	Single weft	y = 1.526x – 1.114 (R^2 = 0.939)	y = 1.331x – 1.528 (R^2 = 0.982)	y = 1.539x – 0.036 (R^2 = 0.954)

Table 7.
Matrices of regression equations and R^2 values between fabric wicking and yarn wicking for twill weave fabric samples.

also aids in the identification of differences in relationship with plain weave samples of similar types as displayed in **Table 6**.

From the tabulated data of twill weave fabrics, the relationship of wicking between fabric and yarns is found to be much stronger (as the range of *R*2 lies within 0.912–0.991) compared with that of the plain weave in the fabric axis of the twill weave creating less amount of disturbance on capillary action due to the lower number of interlacements of twill design. As a result, the smooth movement of liquid within the yarn structure helps in a more effective wicking effect than plain weave structures, strengthening the relation between yarn and fabric wicking. As a result, regardless of the different directions of experimental fabric samples used in this investigation, changes in yarn fineness have a comparatively smaller impact on the connection [44].

7. Conclusion

The cotton fabrics presented in this chapter are only small fractions, which are produced globally for several applications. Absorbency and wicking are the key factors while producing or engineering a fabric for particular end uses such as sports, medical and other technical applications. Therefore, wickability of yarn and fabric has been studied critically as it is completely governed by their structure primarily. Therefore, even minor differences in structure can affect wicking height in both yarn and fabric. Increased fineness diminishes wickability in both ply and single yarn structures separately, but when employed in fabric structures, increased yarn fineness improves wicking [44]. Based on scientific experiments, uniquely presented correlation matrices have expressed the interrelation between cotton fabric and its constituent yarns showing strong correlation but varying nature due to differences in structure, direction of measurement, etc. Hence, these correlations can be used as a quick predictor of cotton fabric wicking based on the measured wicking of their constituent yarns for future research. Further research includes the study of absorbency and wicking with various solvents, temperature, and humidity for wide range of textile materials using these correlations with some modified equations.

Acknowledgements

The authors gratefully acknowledge the assistance provided by Prof. Sumanta Bhattacharya, Former Principal, and Prof. Biswapati Chatterjee, Former Head of the Department in Textile Technology, Government College of Engineering and Textile Technology, Serampore, West Bengal, India.

Author details

Palash Mallick[1*] and Susanta Sekhar De[2]

1 Department of Jute and Fibre Technology, University of Calcutta, Kolkata, West Bengal, India

2 Department of Textile Technology, Government College of Engineering and Textile Technology, Serampore, West Bengal, India

*Address all correspondence to: palashmallick91@gmail.com

References

[1] Hsieh YL, Yu B, Hartzell MM. Liquid wetting, transport and retention properties of fibrous assemblies: Part II: Water wetting and retention of 100% and blended woven fabrics. Textile Research Journal. 1992;**62**(12):697-704. DOI: 10.1177/004051759206201201

[2] Ito H, Muraoka Y. Water transport along textile fibers as measured by an electrical capacitance technique. Textile Research Journal. 1993;**63**(2):414-420. DOI: 10.1177/004051759306300706

[3] Liu T. Coupled mechanical and liquid moisture transfer behaviour of textile materials [PhD thesis]. The Hong Kong Polytechnic University; 2008

[4] Banu KP. Wicking behaviour of fibres, yarns and fabrics [PhD thesis]. Mother Teresa Women's University; 2013

[5] Washburn EW. The dynamics of capillary flow. Physical Review. 1921;**17**: 273-283

[6] Pan N, Gibson P, editors. Thermal and Moisture Transport in Fibrous Materials. Cambridge, England: Woodhead Publishing Limited; 2006. pp. 136-153

[7] Hollies NRS, Kaessinger MM, Bogaty H. Water transport mechanisms in textile materials1 Part I: The role of yarn roughness in capillary-type penetration. Textile Research Journal. 1956;**26**:829-835. DOI: 10.1177/004051755602601102

[8] Hollies NRS, Kaessinger MM, Watson BS, Bogaty H. Water transport mechanisms in textile materials: Part II: Capillary-type penetration in yarns and fabrics. Textile Research Journal. 1957;**27**: 8-13. DOI: 10.1177/004051755702700102

[9] Perwuelz A, Casetta M, Caze C. Liquid organisation during capillary rise in yarns—Influence of yarn torsion. Polymer Testing. 2001;**20**:553-561. DOI: 10.1016/S0142-9418(00)00075-1

[10] Liu T, Choi K, Li Y. Wicking in twisted yarns. Journal of Colloid and Interface Science. 2008;**318**:124-139

[11] Rajagopalan D, Aneja A. Modelling capillary flow in complex geometries. Textile Research Journal. 2001;**71**(9): 813-821. DOI: 10.1177/004051750107100911

[12] Subramanian SN, Venkatachalam A, Subramaiam V. Wicking behaviour of regular ring, jet ring-spun and other types of compact yarns. Indian Journal of Fibre and Textile Research. 2007;**32**: 158-162

[13] Wang N, Zha A, Wang J. Study on the wicking property of polyester filament yarns. Fibers and Polymers. 2008;**9**(1):97-100. DOI: 10.1007/s12221-008-0016-2

[14] De SS, Mallick P. Study the wicking characteristics of cotton yarns. In: 3rd National Conference on Emerging Trends in Textiles, Fibre and Apparel Engineering. Berhampore, India: Department of Textile Technology, Government College of Engineering and Textile Technology; 2016. pp. 8-12

[15] De SS, Porel S, Patra N, Kundu B, Mallick P. Influence of different cotton yarn structures on wicking characteristics. American International Journal of Research in Sciences, Technology, Engineering and Mathematics. 2016;**15**(2):173-176

[16] Das A, Ishtiaque SM, Yadav P. Properties of woven fabrics containing

core-sheath DREF-III yarn in weft. Indian Journal of Fibre and Textile Research. 2003;**28**:260-264

[17] Montgomery DC. Design & Analysis of Experiments. 8th ed. Vol. 5. Singapore: John Wiley & Sons; 2004. p. 427

[18] Walker Esq EM, FTI, Doyle PJ. Some fundamental properties of hosiery yarns and their relation to the mechanical characteristics of knitted fabrics. Journal of the Textile Institute. 1952;**43**(1):19-35. DOI: 10.1080/19447015208663954

[19] Merati AA, Okamura M. Hollow yarn in friction spinning: part I: tensile properties of hollow yarn. Textile Research Journal. 2000;**70**(12):1070-1076. DOI: 10.1177/004051750007001206

[20] Slater K. Comfort properties of textile. Textile Progress. 1977;**9**(4):1-70. DOI: 10.1080/00045167.1977.10750095

[21] Knapton JJF, Truter EV, Aziz AKMA. The geometry, dimensional properties, and stabilization of the cotton plain-jersey structure. Journal of the Textile Institute. 1975;**66**(12): 413-419. DOI: 10.1080/00405007508630536

[22] Postle R. A geometrical assessment of the thickness and bulk density of weft-knitted fabrics. Journal of the Textile Institute. 1974;**65**(4):155-163. DOI: 10.1080/00405007408630442

[23] Lord PR. Handbook of Yarn Production. Cambridge: Woodhead Publishing Ltd.; 2003

[24] Yoo HS, Hu YS, E A K. Effects of heat and moisture transport in fabrics and garments determined with a vertical plate sweating skin model. Textile Research Journal. 2000;**70**(6):542-549. DOI: 10.1177/004051750007000612

[25] Ansari N, Haghighat Kish M. The wicking of water in yarn as measured by an electrical resistance technique. Journal of the Textile Institute. 2000;**91** (3):410-419. DOI: 10.1080/00405000008659517

[26] Das A, Ishtiaque SM. Comfort characteristics of fabrics containing twist-less and hollow fibrous assemblies in weft. Journal of Textile and Apparel, Technology and Management. 2004;**3** (4):1-7

[27] Anbahan Ariadurai S, Prasad P, Whyte IL. Modeling the in-plane permeability of woven geotextiles. Textile Research Journal. 1999;**69**(5):345-351. DOI: 10.1177/004051759906900507

[28] Kissa E. Wetting and wicking. Textile Research Journal. 1996;**66**(10):660-668. DOI: 10.1177/004051759606601008

[29] Roy MD, Chattopadhayay R, Sinha SK. Wicking performance of profiled fibre. Part B: Assessment of fabrics. Journal of The Institution of Engineers (India): Series E. 2018;**99**:1-8

[30] Ramesh Babu V, Ramakrishnan G, Subramaniam VS, Lakshmi K. Influence of the weave factor on the character of fabric wicking measured by a multiple probe vertical wicking tester. Fibres & Textiles in Eastern Europe. 2011;**19** (5,88):60-63

[31] Chattopadhyay R. One Day Seminar on Comfort in Textile. Delhi: Department of Textile Technology, IIT; 2004. p. 20

[32] Sharma N, Kumar P, Bhatia D, Sinha SK. Moisture management behaviour of knitted fabric from structurally modified ring and vortex spun yarn. Journal of The Institution of Engineers (India): Series E. 2016;**97**: 123-129

[33] Mallick P, De SS. Study the wicking phenomena in different directions of cotton plain and twill fabrics made of variable yarn counts. In: International Conference in Textile and Clothing-Present and Future Trends. Kolkata, India: Department of Jute and Fibre Technology, University of Calcutta; 2017. pp. 75-79

[34] Zhu G, Militky J, Wang Y, Sunderlal BV, Kremenokova D. Study on the wicking property of cotton fabric. Fibres & Textiles in Eastern Europe. 2015;**23**(2(110)):137-140

[35] Ozdemir H. Permeability and wicking properties of modal and lyocell woven fabrics used for clothing. Journal of Engineered Fibers and Fabrics. 2017; **12**(1):12-21. DOI: 10.1177/155892501701200102

[36] Bhargava GS. Effect of kind of terry weave on strength, water absorbency and wettability of terry fabrics. Indian Textile Journal. 1985;**95**(9):95-101

[37] Anand SC, Higgins L. Textile materials and products for activewear and sportwear. In: 2nd International Conference of NISTI, New Delhi, India. 2–3 December 2004. p. 204

[38] Silva APD, Anand SC. Responsive garments for sportswear. In: Proceedings on Smart Textile, NIFT, New Delhi, India. 29 November 2000. p. 32

[39] Ishtiaque SM. One Day Seminar on Automotive Textiles, SASMIRA, Mumbai, India. 19 February 2000. p. 67

[40] Patnaik A, Rengasamy RS, Kothari VK, Ghosh A. Wetting and Wicking in Fibrous Materials, Textile Progress. Manchester: The Textile Institute; 2010

[41] Sanyal T. Jute geotextiles: Its types and functions. In: Jute Geotextiles and their Applications in Civil Engineering. Singapore: Springer; 2017. pp. 19-31. DOI: 10.1007/978-981-10-1932-6_3

[42] Erdumlu N, Saricam C. Wicking and drying properties of conventional ring and vortex-spun cotton yarns and fabrics. Journal of the Textile Institute. 2013;**104**(12):1284-1291. DOI: 10.1080/00405000.2013.799258

[43] Wong KK, Tao XM, Yuen CWM, Yeung KW. Wicking properties of linen treated with low temperature plasma. Textile Research Journal. 2001;**71**(1): 49-56. DOI: 10.1177/004051750107100108

[44] Mallick P, De SS. Study the wicking phenomena of cotton woven fabrics and its constituent yarns: Relation between fabric wicking and yarn wicking. Journal of Natural Fibers. 2021. DOI: 10.1080/15440478.2021.1875371

Chapter 7

Extraction, Applications and Characterization of Plant Fibers

Richard Ntenga, Saidjo Saidjo, Annie Wakata, Pagore Djoda, Martin Tango and Etienne Mfoumou

Abstract

During the second half of the twentieth century, industrial and scientific interests in plant fibers (PFs) have resulted in their resounding comeback as engineering materials. This chapter is concerned with the characterization of PF materials. Good knowledge of the properties of these materials is essential for safe design of the related structures. Bast fibers that are collected from the phloem surrounding the stem of certain dicotyledonous plants, for instance, are among the most used, owing to their higher tensile strength. However, for an optimum utilization of PFs, a relevant assessment of their physico-chemical and mechanical properties is very crucial. As it is now well established, PFs' properties are largely influenced by their hierarchic composite microstructure and their viscoelastic behavior. This book chapter focuses on the presentation of various experimental approaches used to characterize the elastic and viscoelastic behaviors of plant fibers. Consideration of their blending in sheet form and relevant mechanical properties will also be of interest.

Keywords: plant fibers, mechanical testing, viscoelastic properties, cell wall, mechanical properties, inverse methods

1. Introduction

In the 1830s, the use of plant resources such as flax, hemp and others was widespread as their fibers were in high demand by the textile, the paper and sailing industries. These plants were grown over large areas for exploitation. However, with the progress of science and technology (loom, steam engine, development of cotton harvesting and processing technique and others), materials such as metals, ceramics, glass, polymers, stones and concrete were preferred to plant resources. In 1900, fiber plants experienced their lowest implantation in terms of surface area. Indeed, the rise of new materials has greatly contributed to the improvement of human living conditions through the construction of more robust and sophisticated habitats, the development of the automotive, railway, aeronautics, textile industries, etc. Subsequently, in a concern for economy, lightness and performance, the development of composite materials was born during the 1930s.

The industrial use of plant fibers began in the early twentieth century with the manufacturing of aircraft seats, fuel tanks or other electronic boxes in plant fibers reinforced polymer materials. The need for securing constructions or structures that are made up of these materials inevitably arose. From an engineering viewpoint, this is taken into consideration during the design, due to a good knowledge of the material characteristics. Plant fibers have specific properties that make them good candidate reinforcing materials for high-performance composites and other applications [1]. However, the mechanical properties of PFs vary considerably both within the same species and from one species to another. Humidity variation, for example, leads to shrinkage or swelling that changes mechanical properties [2]. Similarly, their thermal properties are by far very different from those of synthetic fibers.

Various studies also indicate that plant fibers exhibit, for example, a very complex anisotropic behavior [3–5]. This anisotropy must be accounted for if a reliable design is to be achieved. Close collaboration between scientific disciplines such as botany, chemistry, biochemistry, molecular structural biology, plant genetics, physics and mechanics allows each of them to make a constructive and complementary contribution. PFs must withstand stresses of all kinds when they are associated with their deriving plants. They are loaded when it comes to supporting the weight of the plant or when it comes to resisting the winds, storms and hurricanes so common in their environment. PFs are diverse, and can all be studied for their use as engineering materials, in order to take benefit of the particular advantages offered by each of them. Meanwhile, their mechanical, physical and chemical characterization can differ between members of the same species and from one species to another. They are most often in the form of bundles (technical fibers) comprising one to twenty elementary fibers. They have a complex hierarchical structure inducing anisotropy and, have great geometric and mechanical variability. Humidity variation, for example, leads to shrinkage or swelling and changes in mechanical properties.

The characterization of material generally involves so-called monotonous tests (tensile, compression, torsion, bending or a combination) according to the load's direction (uniaxial or multiaxial), cyclic tests, hardness and resilience tests. Tensile test is undoubtedly the most common test applied to PFs [6–12] because it allows obtaining Young's modulus, strength and elongation at break. Recent works show that PFs exhibit a delayed behavior over time and temperature [13, 14] highlighting their viscoelastic nature. A tensile test alone is therefore not sufficient to characterize these materials.

This chapter is structured in four sections. Following this introduction, Section 2 will give an overview of some essential applications, the supply chain and the techniques of separating fibers from their plant. In Section 3, we will describe the experimental characterization methods generally used to derive their structure morphology and their elastic, viscoelastic and thermomechanical properties. Section 4 is the conclusion.

2. Plant fibers

2.1 Common applications

The natural fiber derived from wood, sisal, hemp, coconut, cotton, kenaf, flax, jute, abaca, banana leaf fibers, bamboo, wheat straw or other fibrous material and the matrix can be a polymeric material. The key advantage of natural fibers and their composites over traditional materials is their biological and environmental durability as well as their superior biodegradability. Natural PFs are increasingly used in several fields of engineering applications because of their interesting properties [15]. Diverse abundance of natural fiber, shapes and forms is caused by their occurrence in different climatic zones, hence stimulating the interest and opportunities to conduct comprehensive studies for identifying new applications for the fibers in industry. Notably, they are gaining popularity due to their optimal use in reinforcement of bio-composite structures. These fibers are biodegradable, structurally sound and environmental friendly. However, a sound theoretical basis for modeling their structure and mechanical behavior has yet to be established. Thus, it will be a priority field of study that will challenge the scientists and researchers.

2.2 Fibers supply chain

The emerging trends and opportunities for natural fibers are broadening due to desirable attributes such as biodegradability, eco-friendly, sustainability and energy efficiency. Sustainability supply chain of natural fibers is assessed and rated based on the following criteria: water usage, CO_2 emissions, cost, availability and any other impacts [16]. Moreover, in the fashion industry, businesses tend to identify the impacts of fibers on brands that contribute to the most impressive reduction in their impact on environmental footprint. Some of the preferred fibers include Linen, Tencel, Bamboo, Recycled Polyester, Recycled Wool, Cork, Organic Cotton and Hemp.

Perhaps the most important factor is the understanding of the entirety of the supply chain of natural fibers and the stages that contribute to having the biggest impacts. Consequently, a map of biodiversity quantitative impact indicators that help the companies determine where to focus their efforts in supply chain management to alleviate natural fiber environmental footprint was developed.

Nowadays, only 23% of companies take into account their environmental footprint when choosing their suppliers and between 40 and 60% of a company's environmental footprint actually comes from its supply chain. Hence, in developing the natural fiber supply strategy, it is critical to understand the role of supply chain management and the associated impacts of environmental footprint. Network analysis, optimization of transhipment costs and decision analysis on optimal solutions to minimize both the supply chain cost and environmental footprint are essential toolkits in the advancement and promotion of natural fibers industry.

Moreover, over the last two decades, the trends in production of plant fibers have been declining due to popularity of synthetic fibers as well as adverse drought conditions. The fiber production plants spread across all continents of the globe. **Table 1** illustrates the trends of different sources of fibers, production capacities and where they are produced.

In 2018, world production of all apparel and textile fibers reached 110 million tons, with natural fiber production estimated at 32 million metric tons. Natural fibers accounted for 29% of the total world fiber production capacity, with most of annual yield variation linked to dry weather conditions. Moreover, the decline in the amounts of natural fibers in total fiber production in the last decade is due to the exponential growth in polyester production, whose demands were triggered by the fast-fashion apparel industry.

2.3 Plant fibers' extraction methods

2.3.1 Methods commonly used

Cellulosic fibers originated from plants and trees such as cotton, flax, hemp, jute, ramie, kapok, coir and bamboo are termed natural PFs. Such fibers are derived from

Fiber source	World production (103 tonnes)	Origin	Country
Abaca	70	Leaf	Malaysia, Uganda, Philippines, Bolivia, Brazil
Bambou	10,000	Stem	Africa, India, Brazil
Banana	200	Stem	Africa, India, Brazil
Broom	Abundant	Stem	
Coir	100	Fruit	India, Sri Lanka, Philippines, Malaysia, Brazil
Cotton Lint	18,500	Stem	India, Europe, USA
Elephant Grass	Abundant	Stem	India, Africa
Flax	810	Stem	Europe
Hemp	215	Stem	Yugoslavia, China
Jute	2500	Stem	India, Egypt, Guyana, Jamaica, Ghana, Malawi, Sudan, Tanzania, Brazil
Kenaf	770	Stem	Iraq, Tanzania, Jamaica
Linseed	Abundant	Fruit	USA
Nettles	Abundant	Stem	Europe
Oil Palm Fruit	Abundant	Fruit	Malaysia, India, Brazil, Indonesia, Philippines
Palmyrah	Abundant	Stem	India
Ramie	100	Stem	Honduras, Mauritius
Roselli	250	Stem	Borneo, Guyana, Malaysia, Sri Lanka, Togo, Indonesia, Tanzania
Rice Husk	Abundant	Fruit/grain	India, Japan, Brazil, Others
Rice Straw	Abundant	Stem	India, Japan, Brazil, Others
Sisal	380	Leaf	East Africa, Bahamas, Antigua, Kenya, Tanzania, India, Brazil
Sun Hemp	70	Stem	Nigeria, Guyana, Siera Leone, India
Wheat Straw	Abundant	Stem	USA, Brazil, India, Canada
Wood	1,750,000	Stem	All Countries

Table 1.
Fiber sources, country and annual production of plant fibers.

various parts of plants including leaves, stems (bast fibers), fruits and seeds. Because all natural PFs are made up of mainly cellulose, they are categorized as 'natural cellulosic fibres', which may consist of one plant cell or an aggregate of cells bounded together by non-cellulose materials. Major commercially used PFs include: seed fibers (cotton, coir, kapok), bast fibers (flax, hemp, ramie, bamboo, banana), leaf fibers (sisal, kenaf, pineapple, abaca). To date, bast fibers are produced and utilized to manufacture a wide array of traditional and novel products including ropes, nets, carpets, mats, brushes, mattresses, paper and board materials. Generally, PFs are classified into two groups, namely soft fibers and hard fibers. Soft fibers are obtained through labour-intensive processes. It involves the following steps: selection of plant and harvesting the plant, partial drying, pounding with stone mallet, scraped with devices similar to comb to clean the fibers, wash the fibers, dry in the sun and finally comb the fibers. Subsequently, the fibers are ready to be spun or twisted into thread or cord. Soft fibers are often used to make ropes, string, nets, bags, and hammocks.

Hard fibers are processed through successive phases of cutting, drying, cleaning, and soaking before they can be woven. They are strong and naturally flexible fibers, thus suitable and utilized to make furniture, birdcages, toys, baskets, and mats.

Figure 1(a) and **(b)** shows the matured flax plants grown under a controlled greenhouse environment and a setup of bench-scale trouph for water retting of flax stems [17].

Historically, most plant fibers were extracted manually, supplemented by natural retting. Evidently, this process is tedious, time-consuming and the extracted quality of fibers depends on the skill of the labourer. Nowadays, these fibers are extracted by chemical, mechanical or biological methods.

Akubueze et al. [18], reviewed the chemical techniques employed to extract fibers from natural plants, which include alkali, acid and other reagents. The typical mechanical extraction methods involve the use of stripping the plant stem (typically known as Bacnis and Leonit processes). The latest mechanical extraction methods utilize the decortication process, whereby the plant stems are crushed between two drum rollers to obtain the fibers after removing the pulp. The use of decorticators increase fiber production by 20–25 times compared with the manual process. With biological processes, both consortium of microorganisms and enzymes are utilized to efficiently extract fibers from plant stems.

Overall, the mechanical extraction is incapable to remove the natural binding material (pectin) from the interspaces of the fibers within fiber bundle, chemical extraction is capable to remove the pectin within the fiber bundle but causes significant environmental pollution, whereas the biological extraction method provides increased fiber yield, with minimum detrimental effects to the environment.

According to the Centre for Learning and Teaching in Art and Design (CLTAD), bast fibers, for example, are generally obtained from the phloem, an inner skin of a plant. These fibers support the cells of the phloem to provide strength to the stem. During processing, the fibers need to be separated from both the interior (xylem) and exterior (epidermis) which is the outermost layer of cells. The processes for separating these fibers from plant stalks are known as retting and decortication. Bast fiber

(a) (b)

Figure 1.
Greenhouse controlled experiments for flax plants [17]. (a) Matured flax plants in Phytotron and (b) matured flax stems undergoing water retting.

bundles are typically several feet long, composed of overlapping cellulose fibers and a cohesive gum (or pectin), which strengthens the stem of the plant. The processes with which the bast fibers are separated significantly influence the quality of fibers as there are many stages involved. Kumar et al. [19], reported that the processing of sustainable fiber starts with fiber extraction and yarn production followed by bleaching, dyeing, softening, printing and drying.

Moreover, the process that separates the fibers into smaller bundles and elementary fibers is known as retting. Fiber retting is a key process and is an important criterion that most industries value because it determines the ultimate properties of the fibers produced. Traditional retting methods include dew and water retting. Dew retting depends on ambient weather conditions, typically takes several weeks and hence the quality of fibers produced varies considerably. Similarly, water retting has been a primary method for low-cost production of bast fibers. The process involves submerging bast straws into water and then the decomposition of the pectic is effected by the activity of anaerobic microorganisms. The quality of retting is assessed by the weight, degumming rate and the fiber properties. The faster rate of weight loss is preferred, the degumming rate is evaluated as the percentage change in pectin content of phloem regions in the raw plant to those in water-retted plant, whereas the desired fiber properties include color, linear density and tensile strength. Ruan et al. [20], reported that water retting improved both whiteness and fineness as well as the mechanical properties of fibers.

Although water retting is capable to produce good quality fibers, the inherent long duration of 7–14 days and associated odor has made it less attractive. The retting period can be reduced to 100 h by using warm water (35°C), but high water consumption and unpleasant odor limit its use to some developing countries. Retting is the process by which pectin gets dissolved or softened from the fiber bundles and separates the fibers from stems through microbial activity. As such, a group of Clostridium microorganism is commonly known to play a significant role in the process by hydrolysing the pectin as it produces pectinase enzyme. These enzymes initially attack the cambium layer and then the other thin-walled cells in the cortex. This phenomenon takes place in most plant bast fibers as they have similar long filament structures, except those from cotton fibers which are single plant cells. As an example, for the retting process conducted in a bench-scale trouph under no-flow process water conditions, there were distinct features on how the fibers separate from bundles. **Figure 2(a)** and **(b)** show the scanning electron microscopy of the unretted and retted fibers of flax.

Figure 3(a) shows that cellulosic fiber production accounted for 6% of the total in 2018, synthetic filament accounted for 45% and synthetic staple 20%. Similarly, **Figure 3(b)** depicts that cotton accounted for 81% of natural fiber production by weight in 2018, jute accounted for 7%, while coir and wool each accounted for 3%.

The synthetic fibers are dominated by polyester, which accounts for nearly 90% of world filament production and 70% of world synthetic staple production. The remaining synthetic fibers are composed mostly of nylon, acrylic and polypropylene.

Perhaps a key factor is to consider the role and contribution of human capital and household social economics. Employment statistics in natural fiber industries is difficult to estimate because households do not engage in consistent annual production. In Ref. [21] it is estimated that about 60 million households worldwide are engaged in natural fiber production, and hence the total employment, reflecting both full-time year-round employment and part-time or seasonal employment, is around 300 million, which represents about 4% of the world's population.

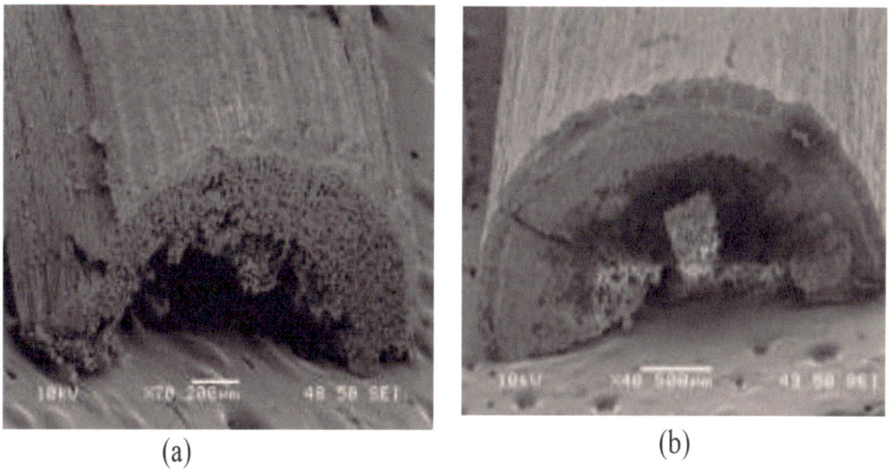

Figure 2.
A SEM shows the microstructure of flax fibers (a) before retting and (b) after the retting process [17].

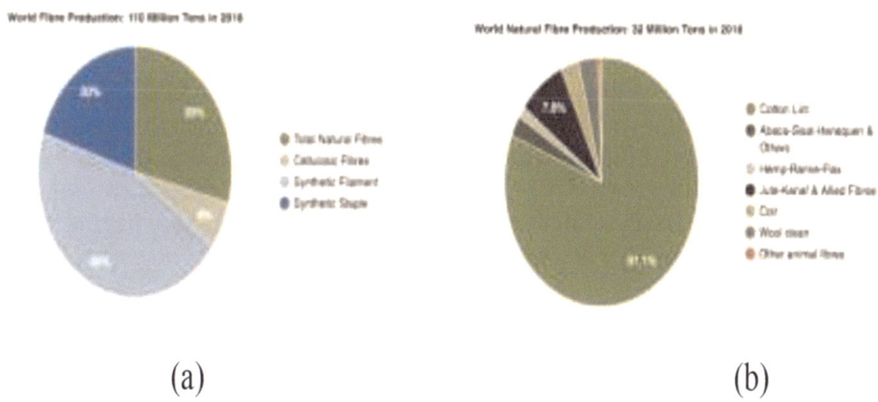

Figure 3.
World total fiber production and natural fiber production [21].

Natural fibers possess superior advantages over synthetic fibers including widespread availability, low cost, low density, moderate strength modulus to weight ratio, high acoustic damping, low manufacturing energy consumption, low carbon footprint and biodegradability. Consequently, there are emerging concerted research initiatives that explore and promote the understanding of the characteristics of natural fibers [15, 22].

2.3.2 Other methods

As discussed in Section 2.3.1 above, dew and water retting are the most common processes for fiber retting. Plant fibers can also be extracted using chemical and enzymatic retting, which provide better control than dew and water retting. Unfortunately, chemical retting while effective in extraction of fibers, causes significant pollution challenges due to higher amount of chemicals utilized. For the chemical extraction methods, alkali and selected reagents have been employed. Alkali treatments promote the fibrillation, whereby the composite fiber bundle is degraded into smaller fibers. Sodium hydroxide (NaOH) is popularly used to reduce the fiber roughness, but also produces good quality fiber. Reagents such as sulfuric acid, hydrogen peroxide, protease and sodium citrate can also be used for chemical extraction [23].

Similarly, enzymatic retting is relatively expensive despite its shorter retting time, yet it produces acceptable fiber quality and is advantageous over other retting processes. In the enzymatic method, the selection of enzymes depends on the type of substrate, composition, size and lignin content. The most common enzymes utilized are cellulases and pectinases. Cellulase enzymes enhance the fiber smoothness by removing fibrils from the outer layer. As such, this results in reduction in the mechanical properties due to the damage caused in the fibers. Pectinases remove the inter-lamellar pectin, which is a natural adhesive compound between fibers.

3. Characterization of plant fibers

3.1 Fiber ultrastructure and morphology

The ultrastructure is about dimensions between the atomic and molecular domains. These are accessed using microscopes. Morphology and quantitative chemistry investigations on plant fibers can be achieved following various analytical techniques such as Fourier transform infrared spectroscopy (FTIR), high-performance liquid chromatography (HPLC) and thermogravimetric analysis (TGA), surface electron microscopy (SEM), atomic force microscopy (AFM) and transmission electron microscopy (TEM) [7, 22]. TEM, which uses the principle of electron diffraction leads to very high magnifications of about 5,000,000. Recent progress in instrumentation has made Raman microscopy an extraordinary analytical tool in biological and plant research [24]. The main advantage of confocal Raman microscopy (CRM) is its lateral spatial resolution and the fact that it provides not only chemical composition information but also structural information.

A plant fiber is a nanostructured, renewable, sustainable and biodegradable composite material (**Figure 4**) [25]. Its cell wall can be likened to a composite lamina, consisting of a few plies reinforced with fibrils. Each individual fiber is composed of a primary wall P and a secondary wall S, itself consisting of three layers S_1, S_2, S_3. In the centre, there may be a cavity called lumen if the cell has not filled up completely during its development. Individual cells are interphased with the middle lamellae (ML) as presented in **Figure 4**. The S_2 layer of the secondary wall represents about 80% of the section and governs the mechanical behavior of the fiber [26]. The middle lamella is a wall 0.5–2 μm thick that surrounds the fiber; it plays the role of matrix that maintains the cohesion of the fibers. It is mainly composed of hemicelluloses, pectin and lignin (about 70%) [27]. **Figure 5(a)–(d)** show micrographs of the RC fiber [28] obtained on a Hitachi H-7650 TEM.

The microfibrillar angle is defined as the angle that the microfibrils form with the longitudinal axis of the cell. These two parameters explain partially the difference in mechanical properties between different types of cortical fibers (**Table 2**) [15]. The microfibrillar angle has a major influence on the elastic properties of plant fibers. The weaker is this angle, the better are the properties for plant fibers to behave as a composite material, which presents better mechanical properties in the reinforcement direction [22, 33]. Xu and Liu [34] predicted that the cell wall elastic modulus of wood varies by a factor of 3 when microfibril angle changes from 40° to 10°.

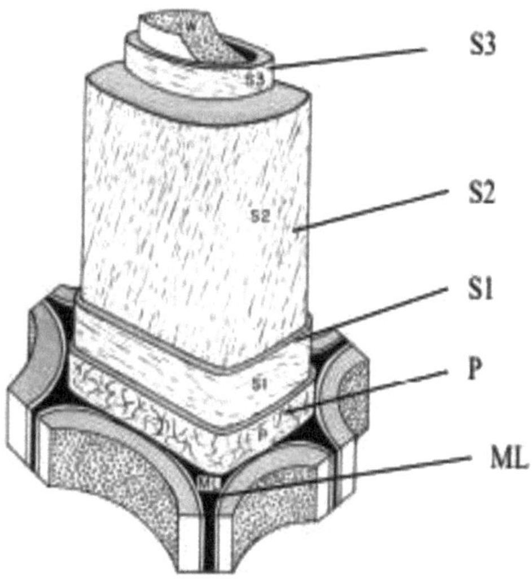

Figure 4.
Simplified structure of the wood cell wall as seen by Coté [25].

The cellulose fibrils are oriented in a helix at an angle called micro-fibril angle, as shown in **Figure 4**. The microfibril angle in the S1 and S3 layers is greater than that of the S2 layer. It means that the fibrils in S1 and S3 layers are almost transversely oriented with respect to the fiber axis. According to the small microfibril angle in the S2 layer, its fibrils are oriented more parallel to the axis of the fiber [35]. In addition, for a given percentage of cellulose, the lower the microfibril angle, the higher the stiffness and strength of the fiber. The greater the microfibril angle, the greater the elongation at break [26]. Each microfibril can be considered as chains of cellulose crystals bound by amorphous zones [36].

The microfibril angle partly explains the elastic deformation of the plant fiber and therefore its elongation at break. Under relatively low tensile forces, a plant fiber undergoes a reversible deformation due to the progressive alignment of cellulose microfibrils with the fiber axis and an elasto-visco-plastic deformation of amorphous polymers. If the stress of the fiber is stronger, the deformation of it enters an irreversible phase that can continue until the rupture. A high microfibril angle implies a greater elastic deformation for a low tensile fiber stress. In addition, there is a negative correlation between the microfibril angle and the corresponding Young's modulus (**Figure 6**) [37].

In order to estimate suitability of different fibers to engineering and other applications, it is necessary, among other things, to determine their mechanical properties in the longitudinal and transverse directions as well as the origin of the viscoelastic properties. Thus, we will present in the following paragraphs a state of the art on the main methods used to evaluate the elastic and viscoelastic properties of PFs. Various methods have been used to measure the angle of microfibrils in the S2 layer, which is generally considered a Z-helix. Nevertheless, some studies using cross-field pit punctuations such as those of Pysznski and Hejnowicz [38] on the tracheids of Norwegian Spruce show that in about 80% of the trees studied, the Z-shaped microfibrils have an angle of 10°–40° while in the remaining 20%, the angle is lower with variations in orientation. A complete list of the different microfibril angle measurement techniques

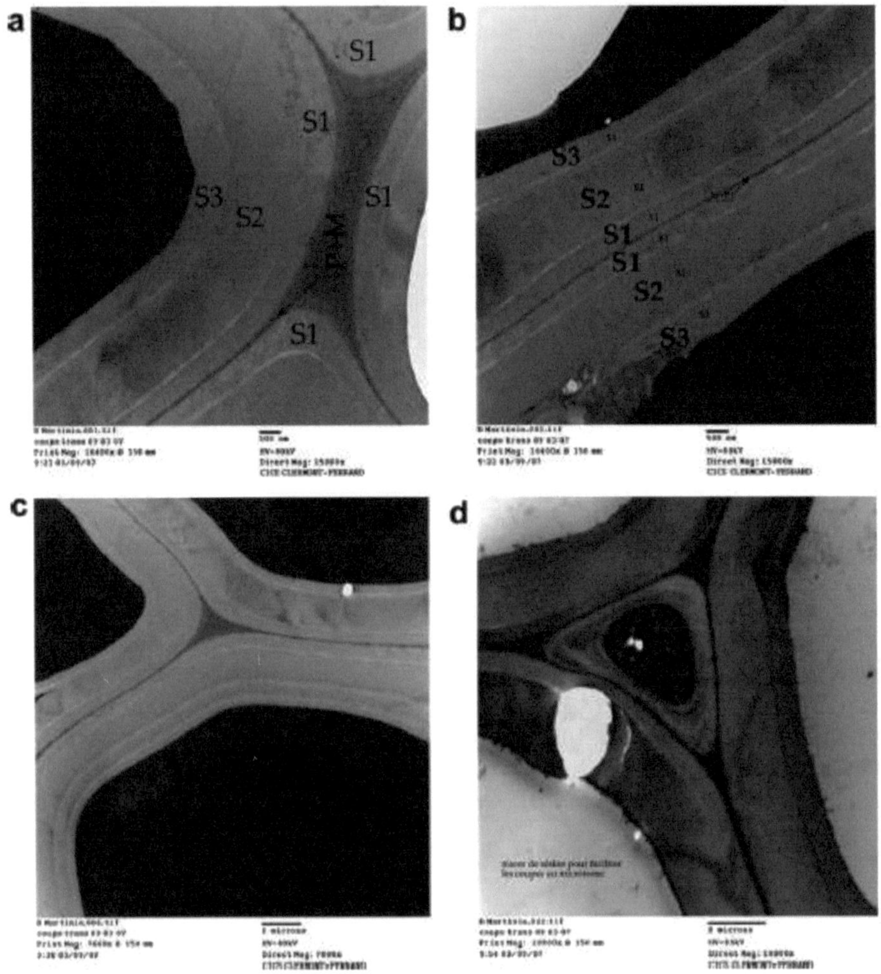

Figure 5.
TEM micrographs of the RC fiber (a) consecutive layers (16,400), (b) layer stacking (16,400), (c) warty sub-layer (7660) and (d) reinforcement by a small cell (10,900) [4].

Fibers	Crystallinity index (CrI)	Microfibril angle (°)	Cross-section area (mm^2)	Length of the cell (mm)	Aspect ratio l/d
Coco		45.0	1.20	3.3	35
Flax		10.0	0.12	2.	1687
Hemp		6.2	0.06	23	960
Jute		8.0	0.12	2.3	110
Ramie		7.5	0.03	154	3500
RC [32]		42	0.05–0.962		>2000
Sisal	56.6–66.2	20.0	1.10	2.2	100

Table 2.
Structural parameters of some plant fibers [29–31].

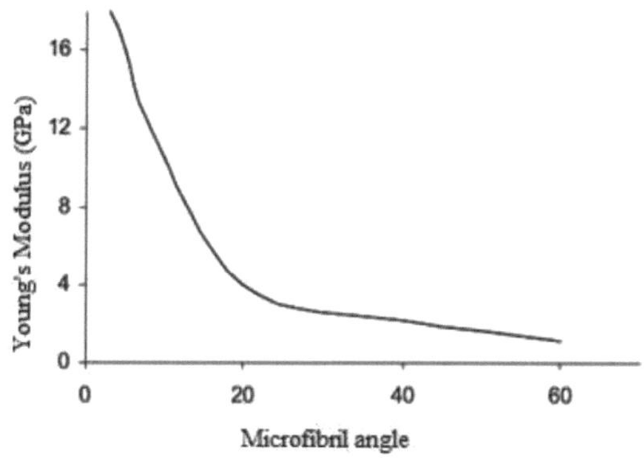

Figure 6.
Variation of the young modulus with the microfibril angle of a unit cell.

with their advantages and disadvantages is given by Huang et al. [39]. Among these techniques, X-ray diffraction is fast, but it is impossible to measure the angle of a single fiber, because of the bundle, only an average of the angle on the X-rays affected cells can be determined. The results obtained by different methods are often contradictory. For example, the work of Herman et al. [40] on individual tracheids shows large variations in the microfibril angle within annual dark circles with a sharp decrease from spring cells to summer cells. While other studies by Lichtenegger et al. [41] using the SAXS (small-angle X-ray scattering) method, on the same cell type shows a higher microfibril angle in summer tracheids than in spring tracheids. Currently, it is necessary to understand where the differences in results obtained by the available measurement methods originate from and to find a method that gives safe and reproductive results. A technique was developed by Jang [42] which uses polarization confocal microscopy based on dichromic cell wall fluorescence when stained with specific fluorochromes showing a high affinity with cellulose. In this technique, sample preparation still needs to be addressed. In fact, very thin samples, only allow observation of fluorescence intensity in the S2 layer without interference with the other layers. A quick but reliable estimate of the Rhectophyllum Camerunense (RC) fiber [28] microfibrils angle was obtained on the SEM (following a microtome longitudinal section of the fiber coinciding with the S2 layer) and fluorescence micrographs.

3.2 Chemical constituents

The chemical composition of plant fibers depends largely on the particular needs of their stemming plant. However, cellulose, hemicellulose and lignin are the main constituents, and their content depends on the age, origin and extraction conditions of the fibers. Cellulose is the chemical constituent that contributes the most to the strength and stability of the plant cell wall and therefore of the fibers. The cellulose content of the fiber largely influences mechanical properties, the economic aspect and the production of the fiber. Fibers with a high cellulose content would be preferable for use in textiles, paper, composites and other fields of activity while those with a high hemicellulose content would be suitable for the production of ethanol and other fermentation products because hemicellulose is easy to hydrolyse in fermentable sugars. Thus, the value of plant fiber and its potential applications depends largely on its cellulose content. Let us say, however, that the value of a plant depends mainly on the quality of its fibers and their end-use and not on the cellulose content itself. As with all-natural products, mechanical and physical properties of natural fibers vary greatly. These properties depend on the chemical and structural composition which depends on the origin of extraction (from leaves, seeds or stems), the local environment where the plants grow, the age of the plants and the climate. The chemical composition, structure, defects and dimensions of the fiber cells are the main parameters that condition all properties of the fibers including mechanical properties [12]. With the exception of cotton, the constituents of plant fibers are cellulose, hemicellulose, lignin, pectin, waxes and water-soluble substances. The average chemical composition of some plant fibers is shown in **Table 3**.

The chemical bonds of the fibers can be determined with FTIR. Crystallographic properties can be analyzed with XRD. TGA, DTA and DSC are used to understand the thermal degradation behavior, the maximum degradation temperature of fibers. Pull-out tests applied to both raw and NaOH treated fibers aim for evaluation of the surface interaction of fibers with polymer matrices for composite materials applications.

In 1838, Anselm Payen proposed that cell walls of many plant cells be made of the same substance to which he gave the name cellulose. Cellulose is a natural polymer whose molecule, formed by long chains, consists of units of D-anhydroglucopyranoses (formula: (C6 H10 O5)n) linked by β-(1,4)-glycosidic bonds in position C1 and C4 (**Figure** 7). It represents the most abundant biological molecule on our planet. It is present in plants, algae, bacteria and some animals.

Cellulose is the major constituent of wood and is the major constituent of cotton and other textile fibers such as flax, hemp, jute and ramie. Its degree of polymerization varies according to the plant species. It can be 14,000 for native cellulose, but the insulation and purification procedures reduce it very sharply by about 2500. Cellulose contributes to the strength and rigidity of the fiber thanks to its strongly oriented chains. These macromolecular chains can be arranged, either regularly, in crystalline

Fiber	Chemical content (%)			
	Cellulose	Hemicellulose	Lignin	Pectin
Abaca	63.2	19.6	5.1	—
Bamboo	48	23	19	—
Cotton	83	5	—	—
Flax	65–70	10–16	2.9	2–4
Hemp	67	16.1	4	—
Jute	55–64	12–18	12–33	0.2
Kenaf	55–59	18–20	6.8–8	4.5–5
Ramie	68.6	13.1	0.6	—
RC	68.2	16	15.6	—
Sisal	54–66	12	7.3	0.8
TJ	62.7	14.5	4.1	7.6
Wood	83	5	19–26	0

Table 3.
Chemical contents of some fibers.

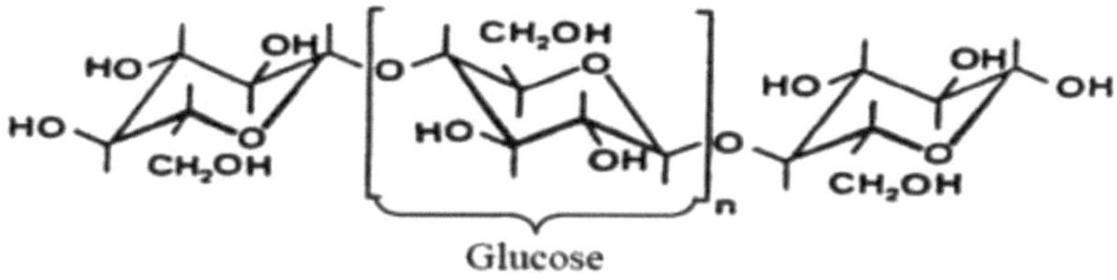

Figure 7.
Cellulose molecule.

regions, or randomly in amorphous regions. Mechanical properties of natural fibers depend on their type of cellulose, as each type has its own cellular geometry. If cellulose is a prime structural constituent for the vast majority of plant cell walls, then hemicellulose with lignin acts as binding materials. Properties depend on the fiber cell geometry of each type of cellulose and its degree of polymerization.

Hemicelluloses represent the second most abundant constituent of plant fiber. Hemicelluloses are polysaccharides found in lignocelluloses alongside cellulose and pectin. Hemicelluloses, unlike cellulose, are composed of several sugars that form short chains with ramifications. The sugars present can be divided into different groups: pentoses (xylose, arabinose), hexoses (glucose, mannose, galactose), hexo-uronic acids (glucuronic acid and methyl-glucuronic acid) and L-deoxyhexoses (rhamnose and fucose). Hemicelluloses are, by definition, water-soluble polysaccharides that can be extracted from the plant cell walls using alkaline solutions. They are the most hydrophilic biopolymers in the cell wall that promote moisture absorption. In their natural state, they have a degree of polymerization that varies from 200 to 300, and their structure depends on the plant species. The best-studied class of hemicelluloses are xyloglucans. They have a bridging role between cellulose microfibrils in order to strengthen the cell wall by interaction with cellulose and, in some cell walls, with lignin. They consist of a glucose chain and short side chains of xylose, galactose and fructose.

Lignin together with cellulose and hemicelluloses is part of the wood industry. Its proportion in wood varies between 15 and 30% [43]. Lignin or 'lignins' are three-dimensional polymers from the radical polymerization of three phenylpropenoic alcohols: coniferryl alcohol, sinapyl alcohol and p-coumaryl alcohol [44]. Lignin contributes to the rigidity of cell walls, and thus to the erect port of terrestrial higher plants. Lignin also offers a protective barrier against the microbial attack of plants. Indeed, due to its chemical nature, lignin is very resistant to various chemical agents and biological degradation. To sum up, lignin polymers make the cell wall rigid and impermeable, allowing the transport of water and nutrients through the vascular system by protecting plants from microbial invasion. Lignin is totally amorphous and hydrophobic. It is not hydrolysed by acids, but hot soluble in soda, easily oxidized and also condensable with phenol.

Pectins are polymers of acidic polysaccharides, composed of a main chain of uronic acid bound in 1–4. Regularly, rhamnose molecules are interspersed between these monomers by bonds 1–2 and 1–4. Some of these rhamnose units carry side chains composed of neutral oses among which galactose and arabinose are the most abundant. The type of bond between uronic acid and rhamnose molecules forms elbows. The pectin macromolecule appears like a zigzag. This arrangement contributes to its special properties and provides some flexibility to plants. Pectins are extracted from the fiber by a chemical method either by boiling water or by ethylene diamine tetraacetic acid.

3.3 Density

Different methods can be used including solid pycnometers or gas pycnometers [45–47]. The choice of gases (helium for example) or immersion liquids such as toluene, ethanol and xylene is decisive for quality results [46, 47]. Fibers must be dried for at least 72 h in a desiccator containing silica (previously regenerated). Fibers are then cut into lengths of 5–15 mm and then introduced into the pycnometer which is eventually placed in the desiccator for at least 24 h. Before carrying out the hydrostatic weighing with the immersion liquids, the vortex agitation of fibers to evacuate the microbubbles between needs to be done. Significant degassing could occur at this stage and provides information on the porosity rate of the fibers [28].

3.4 Mechanical characterization

3.4.1 General considerations

In general, PFs are suitable for reinforcing plastics (thermosets and thermoplastics) and textiles manufacturing thanks to their relatively high strength and low density. The tensile strength and the modulus of elasticity of PFs are very important characteristics for the use of fibers as reinforcements in composite and textile materials. However, the tensile test data for most fibers in service have yet to be studied, as the data found in the literature are scattered and often unreliable. In fact, methods used for the characterization are not identical. **Table 4** shows the tensile mechanical properties of some plant fibers compared to synthetic fibers [48]. The properties of the fibers and their structure depend on several factors such as the origin, variety, conditions of growth and harvesting of fibers associated with the treatments, the location in the stem, the presence or absence of a lumen, measurement techniques that vary greatly from one research team to another. These factors can make a difference for the same type of fiber and influence test results.

Selection of plant fiber implies a prior study of its mechanical properties, chemical resistance, dimensional stability, separation process, etc. It is worth recalling that linear cellulosic macromolecules are linked by hydrogen bonds and are closely associated with hemicelluloses and lignin, which confer stiffness to fiber. One of the issues of natural fibers is the scattered information and the differences in mechanical properties reported. Likewise, the lack of standards for both producers and users of these materials regarding methods to collect, process, post-process and characterize plant fibers underlines the complexity in the selection.

3.4.2 Quasi-static tensile test

Quasi-static tensile test is the method commonly used in the literature for the characterization of the mechanical properties of plant fibers in the longitudinal direction. This type of characterization presents challenges linked to the assembly and to the single nature of the fiber. In addition, the geometry of the plant fiber makes it often difficult to conduct the tests. Therefore, evaluation of the mean diameter along the fiber using a microscope is necessary for the performance of the test. The single fiber is mounted on a paper frame and a drop of glue is used to stick the fibers. The role of this paper frame is to facilitate the handling and alignment of the fiber on the jaws of the experimental device as shown in **Figure 8** [32].

Fiber	Density (g/cm^3)	Diametre (μm)	Length (mm)	Tensile strength (MPa)	Young modulus (GPa)	Elongation at break (%)	Moisture content (%)
Abaca	1.5	10–30 (20)	4.6–5.2 (4.9)	430–813 (621.5)	31.1–33.6 (32.35)	2.9	14
Bamboo	0.6–1.1 (0.85)	25–88 (56.5)	1.5–4 (2.75)	270–862 (566)	17–89 (53)	1.3–8 (4.65)	11–17 (14)
Banana	1.35	12–30 (21)	0.4–0.9 (0.65)	529–914 (721.5)	27–32 (29.5)	5–6 (5.5)	10–11 (10.5)
Coir	1.2	7–30 (18.5)	0.3–3 (1.65)	175	6	15–25 (20)	10
Cotton	1.21	12–35 (23.5)	15–56 (35.5)	287–597 (442)	6–10 (8)	2–10 (6)	33–34 (33.5)
Flax	1.38	5–38 (21.5)	10–65 (37.5)	343–1035 (689)	50–70 (60)	1.2–3 (2.1)	7
Hemp	1.47	10–51 (30.5)	5–55 (30)	580–1110 (845)	30–60 (45)	1.6–4.5 (3.05)	8
Jute	1.23	5–25 (15)	0.8–6 (3.4)	187–773 (480)	20–55 (37.5)	1.5–3.1 (2.3)	12
Kenaf	1.2	12–36 (24)	1.4–11 (6.2)	295–930 (612.5)	22–60 (41)	2.7–6.9 (4.8)	) 6.2–12 (9.1)
Pineapple	1.5	8–41 (24.5)	3–8 (5.5)	170–1627 (898.5)	60–82 (71)	1–3 (2)	14
Ramie	1.44	18–80 (49)	40–250 (145)	400–938 (669)	61.4–128 (94.7)	2–4 (3)	12–17 (14.5)
RC	0.94	70–350 (120)	—	450–1500 (557.1)	5.8 (±3.5)	27.5	—
Sisal	1.2	7–47 (27)	0.8–8 (4.4)	507–855 (681)	9–22 (15.5)	1.9–3 (2.45)	11
TJ	(1.398)	40–90 ()		(404.0)	(32.3)	(1.8)	

Table 4.
Mechanical properties of some selected plant fibers versus synthetic fibers [48].

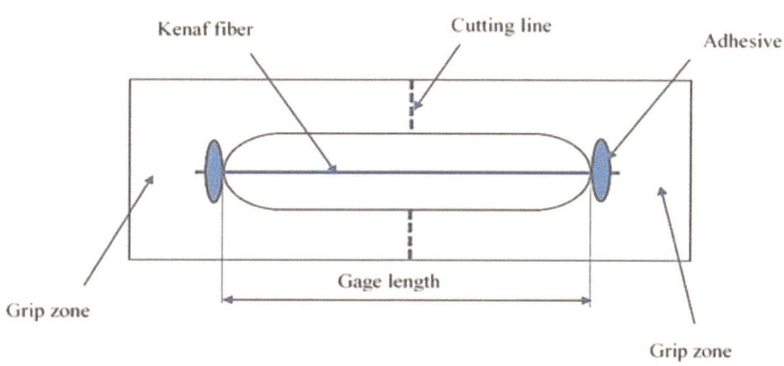

Figure 8.
Tensile test and gripping tab specimens for plant fibers.

The large dispersion of the mechanical properties of the plant fibers observed (**Figure 9**) is mostly related to the test conditions. The research work by Ntenga et al. [14] focused on the choice of the stress speed and the gage length, in order to keep the deformation in the elastic domain and reduce this dispersion during the tests. The machine cross-head speed of 1 mm/min and the gage length of 10 mm were found to cause less dispersion of the mechanical properties in a tensile test.

3.4.3 Nano-indentation test

Nanoindentation is a technique used to characterize the longitudinal and transverse mechanical properties of fibers at the cell wall scale. Commonly measured properties are Young's modulus and material hardness. In the literature, nanoindentation tests have been carried out to access both transverse and longitudinal mechanical properties on wood fibers [31, 49] and recently on flax fibers [50]. According to Cisse [51], nanoindentation only gives access to local behavior of the fiber, and the identification of mechanical properties requires knowledge and use of a behavior model. The testing technique consists of applying a force to the indenter and taking the area of the indentation, in order to determine the Young's modulus and the hardness of the material (**Figure 10(a)** and **(b)**).

A typical set of nanoindentation tests results [53] is shown in **Figure 11**.

Differences in transverse and longitudinal modulus noted between the fibers can be explained not only by the differences in micro-fibrillary angles but also by the rate of cellulose that varies between fibers. Hemp and sisal in particular have a cellulose content of around 60%, while that of flax is over 75%; however, the mechanical properties of cellulose are much superior to those of lignin, hemicelluloses and pectins, other constituents of natural fibers [50].

3.4.4 Dynamic mechanical analysis

A large amount of work exists in the field of vibration-based non-destructive testing (NDT) including an extensive survey of over 300 papers by Kong et al. [54]. Indeed, the vibration-based technique has been a very active area of research for many years, however, has always dealt with rigid bodies. As an extension of the use of

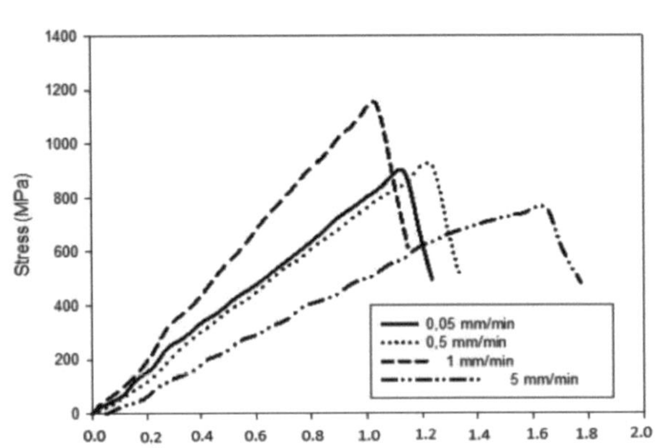

Figure 9.
Tensile stress/strain curves for the four cross-head speeds of gage length 10 mm [14].

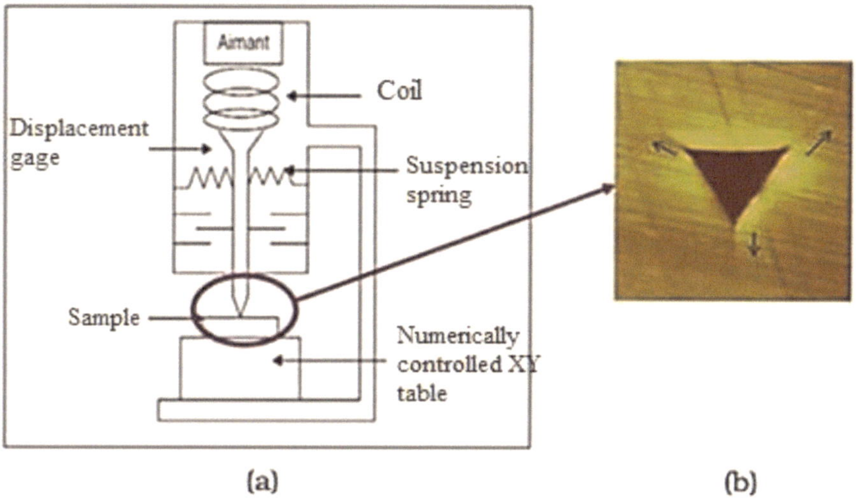

Figure 10.
(a) Nano indentation experimental device and (b) indentor impression Berkovich [52].

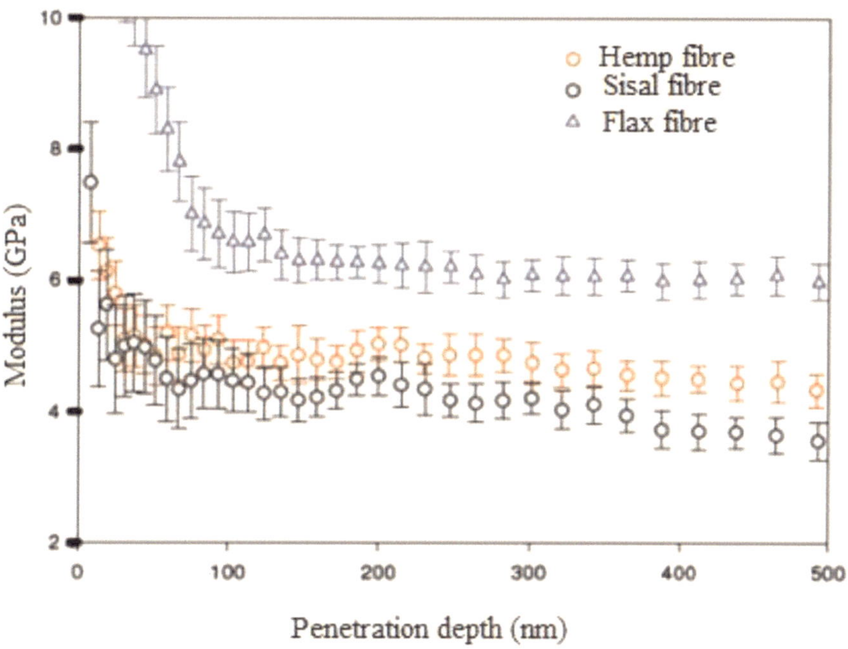

Figure 11.
Transverse modulus of plant fibers obtained in nano indentation.

this technique, the purpose of this section is to present the applicability of the low-frequency vibration-based technique towards estimation of dynamic Young's modulus of natural fiber-based materials, initially having no bending stiffness. This technique enhances the applicability of non-contact acoustic non-destructive testing to the estimation of dynamic characteristics of thin materials, where the current standard method [55] is not applicable.

Let us consider a thin rectangular specimen having a length b, a width a, a thickness h and a density ρ. **Figure 12** shows the specimen configuration in a Cartesian coordinate system at equilibrium (i) and vibrating in flexural mode (ii) and (iii).

The specimen, considered as a membrane, initially has no bending stiffness. It is then slightly stretched in the y-direction, in order to make it possible to vibrate transversally (i.e. in the z-direction). The tensile force F is assumed to remain constant during small vibrations in the y-z plane.

In general, for a specimen having intrinsic elasticity, the equation of motion is expressed as follows:

$$\frac{\partial^2 \xi}{\partial t^2} - c^2\left(\frac{\partial^2 \xi}{\partial x^2} + \frac{\partial^2 \xi}{\partial t y^2}\right) + d^2\left(\frac{\partial^2 \xi}{\partial x^2} + \frac{\partial^2 \xi}{\partial t y^2}\right)^2 \xi = \frac{p(x, y, t)}{\rho h} \tag{1}$$

where ξ is the displacement normal to the plane (x,y) which coincides with the equilibrium position of the membrane, c is the velocity of propagation of bending waves, p is the external pressure on the surface of the membrane. c and d are defined as follow:

$$c = \sqrt{\frac{T}{\rho h}} \text{ and } d^2 = \frac{Eh^2}{12\rho(1-\nu^2)} \tag{2}$$

where E, T and ν are the elastic modulus, tensile force per unit length of the edge, and Poisson's ratio.

The frequency equation with the fixed-fixed boundary condition shown in **Figure 12** above was derived in Mfoumou et al. [56] to obtain the frequency of vibration ω_{mn} of each bending mode as follows:

$$\omega_{mn} = c\sqrt{\left(\frac{\pi m}{a}\right)^2 + \left(\frac{\pi n}{b}\right)^2}\left\{1 + \frac{d^2}{2c^2}\left[\left(\frac{\pi m}{a}\right)^2 + \left(\frac{\pi n}{b}\right)^2\right]\right\}, m, n = 0, 1, 2, \ldots \tag{3}$$

where m and n are the mode numbers.

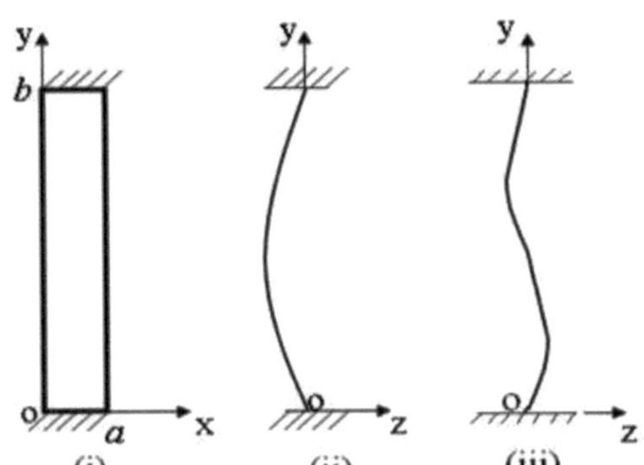

Figure 12.
Specimen configuration (i): undeformed, and vibrating at (ii): fundamental frequency, (iii): second frequency in flexural mode.

DOI: http://dx.doi.org/10.5772/intechopen.103093

For a plant fiber-based material considered as a membrane; therefore, no account of intrinsic elasticity is taken so that Eq. (3) is simplified, and the normal frequencies equation is expressed as:

$$\omega_{mn} = c\sqrt{\left(\frac{\pi m}{a}\right)^2 + \left(\frac{\pi n}{b}\right)^2}, m, n = 0, 1, 2, \ldots \tag{4}$$

The Young's modulus can therefore be determined using the flexural resonance method by monitoring normal modes of vibration. These modes for an oscillating system are special solutions where all the parts of the system are oscillating with the same frequency. At these modes, considering only bending modes in the length direction ($m = 0$), the relationship between the frequency in hertz and the state of strain was established as follows [56]:

$$f_{0n}^2 = \frac{E.n^2}{4.\rho.b^2}.\varepsilon \tag{5}$$

thus, enabling extraction of the constant E from experimentally measured normal mode frequencies and corresponding strains.

3.4.5 Creep test

Both creep experiment and relaxation experiment are two techniques commonly used to characterize the delayed behavior of 'conventional' materials. A creep test consists of imposing an almost instantaneous stress load on the plant fiber and maintaining it constantly over time and then proceeding to a discharge. The resulting deformation under the action of the load is creep, and that under the action of discharge is recovery. In general, the creep responses can be broken down into three stages depending on the strain rate as shown in the following **Figure 13**. The first stage in which creep occurs at a decreasing rate is called primary creep; the second step, commonly called secondary creep, is carried out at a relatively constant speed; and the third stage, tertiary creep, occurs at an increasing rate and terminates with material fracture.

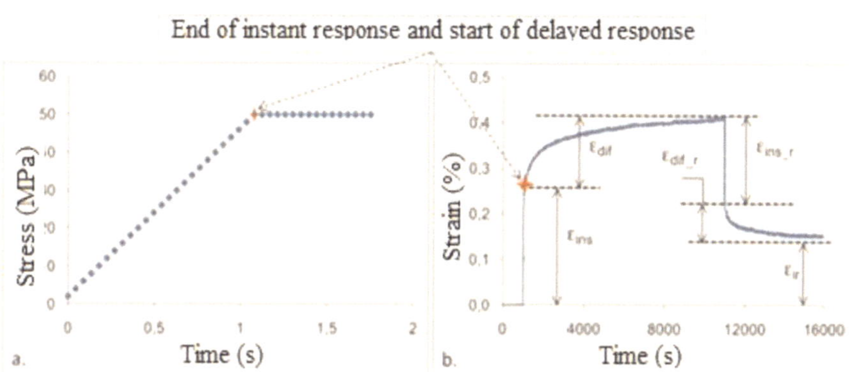

Figure 13.
Creep/recovery test of an elementary hemp fiber under a constant environment [53].

The creep test was successfully carried out on an elementary hemp fiber and the results allowed it possible to highlight the viscoelastic nature of the plant fiber [51]. **Figure 13** shows the creep test results obtained.

3.4.6 Relaxation test

3.4.6.1 The context

When a constant strain is applied to a material for a long period, cross-links or the primary bonds that form between molecules start breaking with time and spontaneously lose their bonding capability. High level of strain or long period is the main reason for intermolecular bond breakage, thus creating stress decay over time, called stress relaxation. The rate of bond breakage influences the rate of stress relaxation. Other factors control the rate of bond breakdown, such as stress on the bond, chemical interference, molecular chain mobility which allows molecular chains to move out from their position. The behavior of stress relaxation in plant fibers is also influenced by temperature, humidity, and strain levels. The stress relaxation tcsts are therefore mainly performed with different ranges of temperature, humidity and strain levels. The time taken to reach the end of relaxation is called relaxation time. From other studies, it is reported that at higher temperature relaxation time becomes shorter, while at lower temperature it becomes longer but the shape of relaxation does not change with temperature [57]; moreover, the variation of strain level affects the stress relaxation [58]. The literature also reports the sensitivity of this class of material to loading-directionality, and ductile and brittle phenomena [59].

3.4.6.2 Stress relaxation measurement

During structural design, the properties of the material must be considered. Elastic Modulus is one of the most important material properties describing the stiffness of the material. When a force is applied to an object, modulus of elasticity or elastic modulus gives the mathematical description of the object's tendency to be deformed elastically.

In orthotropic materials such as wood-based natural fibers, the strain quickly increases linearly with the stress, then exhibit a nonlinear behavior when the strain exceeds the proportional limits. When the stress relaxation tests are conducted for a very small deformation, the viscoelasticity of the material can be considered linear. During stress relaxation test, the material relieves stress over time as well as the elastic modulus of material $E(t)$ also decreases with time at a constant temperature. According to linear viscoelastic material [60], the elastic modulus relaxation can then be defined as:

$$E(t) = \frac{a(t)}{s_o} \tag{6}$$

where, s_o is the constant strain and $a(t)$ is stress of material as a function of time. Indeed, elastic modulus relaxation is the relaxation of modulus of elasticity of material.

3.4.6.3 Sample, experiments, and results

A rectangular strip of specimen is placed between the clamps of the tensile test machine (see **Figure 8**), and it is slightly loaded within its elastic region. The specimen is tested in uniaxial stress-state at a strain rate of 1 mm/mm with 0.4% strain changes.

The elongation is kept constant at 0.4% strain level (1 mm extension) for 5400 s and time, stress, and strain are recorded.

Experiments were carried out for paperboard (PPR) without crack and PPR with crack. Five specimens were tested for each case and each experiment continued for 5400 s (1.5 h) with 1 mm extension. The reason for taking 1 mm extension was to keep the deformation within the elastic region.

The stress relaxation of each specimen was monitored and analyzed at constant.

elongation. The load, stress and time data for constant strain were obtained from the experiments. From the testing of five specimens in each case, we have plotted stress versus time curves. The plotted stress relaxation of PPR without and with the presence of a side crack is presented in **Figure 14**.

Figure 15 show the stress relaxation behavior of PPR at different strain levels (two different extension levels, 1 mm and 0.5 mm).

3.4.6.4 Formulation of relaxation

The data obtained from the stress relaxation experiments are decreasing type of data with function of time and this type of data can be fitted to the poly-exponential function of the following form:

$$y(t) = \sum_{i=1}^{N} a_i e^{-b_i t} \tag{7}$$

where, a_i and b_i are the unknown parameters. Several methods have been developed to estimate these parameters. The most available methods are graphical method, regression-difference equation method, method of partial sums, Fourier Transform method, Foss's method [61] for a sum of two exponentials. However, their uses are limited. For example, the graphical method is not suited where there are consistent fluctuations, regression-difference method and method of partial sums are only appropriate for equally spaced data and Fourier Transform method is suitable for

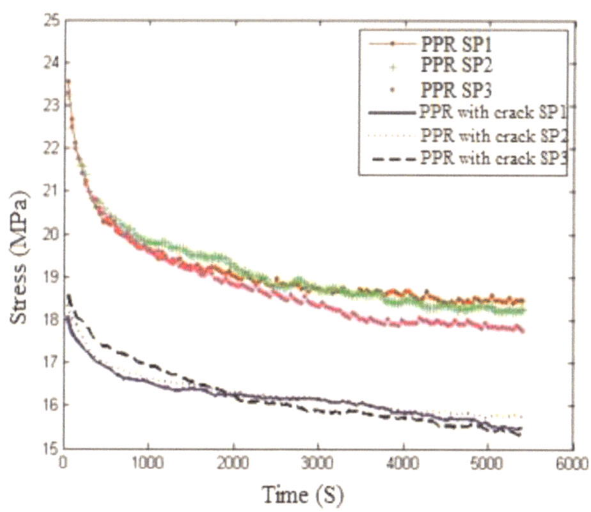

Figure 14.
Stress relaxation of paperboard with and without crack.

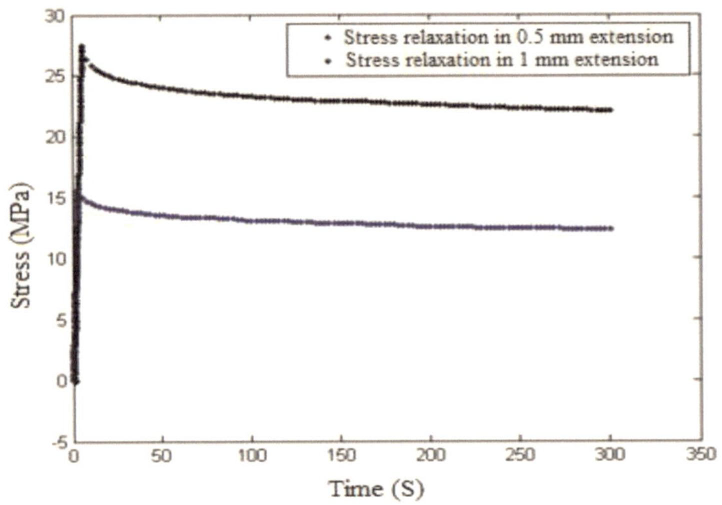

Figure 15.
Stress relaxation of paperboard for 1 mm and 0.5 mm extension.

exponentially spaced observation. On the other hand, the use of Foss method is broader than any other method, and even not equally and exponentially spaced data can be treated using this method [61].

3.4.6.5 Model parameters extraction using experimental data

The parameters of a set of mechanical models can be calculated from experimental data. MATLAB, for example, can be used to extract the parameters from the data. To analyze the suitability of the mechanical model with the experimental stress relaxation, Maxwell Model, Two-unit Maxwell Model, Modified Two-unit Maxwell Model, Standard linear solid model are constructed and then compared with the experimental relaxation. Analytical description of these models is given in [62].

In Ref. [56] we have chosen Foss method to develop curve fitting for all models and then compared with the experimental relaxation. Whereas in Ref. [15] we used the Zapas-Phillips method. The best-fitted model with the experimental data was then selected to analysis all experimental data and mathematically stress relaxation equations were derived.

To predict the stress relaxation behavior of natural fibers, we derived the mathematical equations for PPR with and without presence of crack. These equations were derived by the Modified Two-unit Maxwell model which suits best with the experimental result. Though we carried out our experimental tests with five specimens for each kind of test and among them three specimen-data were taken into consideration, but here we will construct the stress relaxation equation for only one specimen for each case.

Below the comparison, diagrams between experimental relaxation data and the Modified Two-unit Maxwell are shown in **Figures 16** and **17**. The stress relaxation equation for each case is derived using Modified Two-unit Maxwell model.

3.4.7 Inverse characterization

Suitability of materials inverse characterization, destructive or non-destructive, is widely investigated [52, 63, 64]. Furtado et al. [65] used an ultrasound shear wave

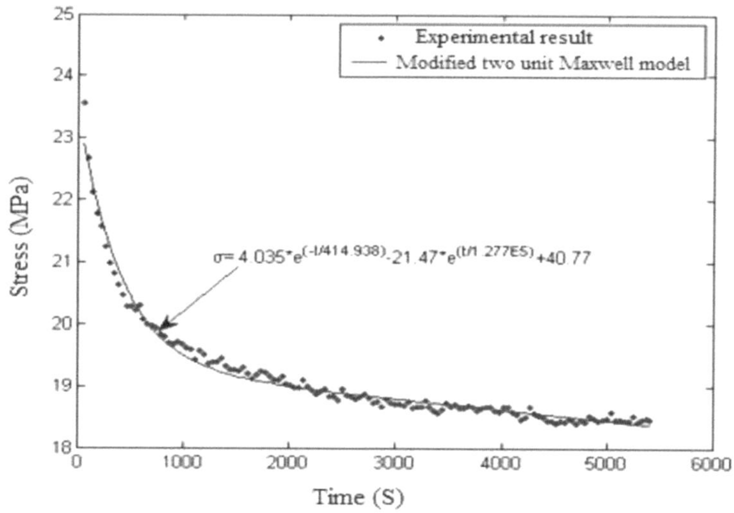

Figure 16.
Stress relaxation of paperboard—curve fitting.

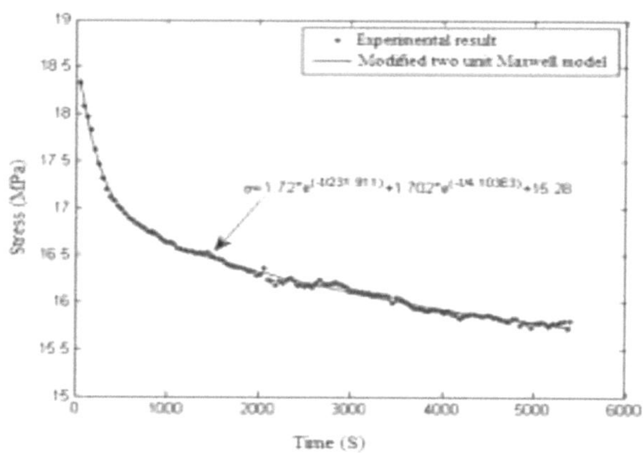

Figure 17.
Stress relaxation of paperboard with crack–curve fitting.

viscoelastography method to determine the viscoelastic complex shear modulus of macroscopically homogeneous tissues. Ilczyszyn et al. [66] performed the mechanical characterization of flax fibers using an inverse optimization simplex method.

The aim here is to use macro-micro approaches to achieve an efficient estimation of the fiber properties. In fact, homogenization laws of the micromechanics of the elastic/viscoelastic behavior of composite materials provide relationships of the properties of these materials in terms of their constituents' properties. For an orthotropic material, the knowledge of its off-axes elastic modules in a set of θ directions leads to the calculation of the fibers' elastic constants for instance. Analytical relationships of elastic constants that account for the orientation of the fibers can be found in the literature. A presentation of an inverse method based on the composite cylinder assembly proposed by Hashin [67] to characterize the anisotropy of plant fibers was discussed in Ref. [64] for a transversely isotropic 2-phase composite. In this case, the matrix and fiber phases are assumed isotropic and transversely isotropic respectively. Using a stress field in cylindrical coordinates and applying Hooke's law gives the following results:

(x, y, z) stand for off-axis Cartesian coordinates (reference coordinate system).

$(1, 2, 3)$ represent material axes of a unidirectional composite.

For a tensile test in the x-direction, the Young modulus is expressed as:

$$\frac{1}{E_{xx}} = \frac{1}{E_{11}} \cos^4\theta + \left(\frac{1}{4G_{23}} + \frac{1}{4K_{23}} + \frac{\nu_{12}^2}{E_{11}}\right) \sin^4\theta + \left(\frac{1}{G_{12}} - 2\frac{\nu_{12}}{E_{11}}\right) \sin^2\theta \cos^2\theta \quad (8)$$

There are five independent properties to be determined $(E_{11}, G_{23}, G_{12}, K_{23}, \nu_{12.})$

Analytical expressions of the five properties in terms of fiber and matrix phase properties and the volume fractions are given by:

$$E_{11} = E_z^f \phi^f + E_z^m \phi^m - \frac{\nu_{z\theta}^f - \nu^m}{\Delta} \left[E_z^f \left(\frac{\nu_{rz}^f}{E_r^f} + \frac{\nu_{z\theta}^f}{E_z^f} \right) - 2\nu^m \right] \phi^m \quad (9)$$

$$\bar{\nu_{12}} = -2\frac{1 - (\nu^m)^2}{E^m} \frac{\nu_{z\theta}^f - \nu^m}{\Delta} + \nu^m \quad (10)$$

$$k_{23} = \frac{E^m \phi^f (I_1 I_4 - I_2 I_3)}{2[-(1 - \nu^m)\phi^f (I_4 L_1 - I_2 L_2) + (1 + \nu^m)\phi^f (I_4 L_1 - I_2 L_2 - I_4 I_1 + I_2 I_3) - \nu^m \phi^m (I_1 L_2 - I_3 L_1)]} \quad (11)$$

$$\frac{\bar{G_{12}}}{G^m} = \frac{G^m \phi^m + G_{rz}^f (\phi^f + 1)}{G^m (\phi^f + 1) + G_{rz}^f \phi^m} \quad (12)$$

$$\frac{\bar{G_{23}}}{G^m} = 1 + \frac{\phi^f}{G^m \left(G_{r\theta}^f - G^m \right) + (k^m + \frac{7}{3} G^m)_{\overline{(2k^m + \frac{8}{3}G^m)G_{rz}^f}} \phi^m} \quad (13)$$

with

$$G_{r\theta}^f = \frac{E_r^f}{2\left(1 + \nu_{r\theta}^f\right)} G^m = \frac{E^m}{2(1 + \nu^m)} k^m = \frac{E^m}{3(1 - 2\nu^m)} \quad (14)$$

$$\Delta = \left[\frac{\nu_{r\theta}^f}{E_r^f} - \frac{1}{E_\theta^f} + \nu_{z\theta}^f \left(\frac{\nu_{rz}^f}{E_r^f} + \frac{\nu_{z\theta}^f}{E_z^f} \right) \right] \frac{\phi^m}{\phi^f} - \frac{\nu^m}{E^m} \frac{\phi^m}{\phi^f} - \frac{1}{E^m} \frac{\phi^f + 1}{\phi^f} + 2\frac{(\nu^m)^2}{E^m}; \quad (15)$$

$$I_1 = \phi^m \left(\frac{\nu_{rz}^f}{E_r^f} + \frac{\nu_{z\theta}^f}{E_z^f} \right) + \frac{2\nu^m}{E_m} \phi^f; I_2 = -\left(\frac{\phi^m}{E_z^f} + \frac{\nu_{z\theta}^f}{E_m} \right) L_1 = -\left(\frac{\nu_{rz}^f}{E_r^f} + \frac{\nu_{z\theta}^f}{E_z^f} \right); \quad (16)$$

$$I_3 = \left(\frac{\nu_{r\theta}^f}{E_r^f} - \frac{\nu_{z\theta}^f}{E_\theta^f} \right) \phi^m - \frac{1}{E_m} (\nu^m - \phi^f - 1); \quad (17)$$

$$I_4 = \frac{\nu_{z\theta}^f}{E_z^f} \phi^m + \frac{\nu^m}{E_m} \phi^f L_2 = -\left[\frac{\nu_{r\theta}^f}{E_r^f} - \frac{1}{E_\theta^f} + \frac{1}{E_m} (\nu^m - 1) \right] \quad (18)$$

Eqs. (8)–(13) are then solved for $E_r^f, E_z^f, \nu_{rz}^f, \nu_{z\theta}^f, G_{r\theta}^f$ *and* G_{rz}^f using a multiparameter optimisation algorithm.

4. Conclusion

There are evolving global challenges on the utilization of non-renewable resources in the manufacturing industry and increasingly stringent environmental legislation. Both consumers and regulatory agencies are thriving for products that reduce dependency on fossil fuels and thus, are more environmentally friendly. As such, this paves for an opportunity to embrace the use of natural fibers in products and composites leading to significant growth of biobased economy, which the present chapter intends to stimulate.

The field of study of plant fibers that can be industrially exploited remains open. In this chapter, a particular emphasis has been put on their production, in particular on the methods that are generally used to separate them from their originating plants. To date, the question of improving the quality of the extracted fiber has been satisfactorily answered, particularly as regards the possibility of combining several methods when necessary. Some other questions still require research. These include, among others, growing conditions for seed multiplication and fiber production, harvesting methods, optimisation of fiber separation, the molecular basis for improving fiber decortication and performance. The knowledge gained from this work could be used to design new varieties of fibers, tailored for specific industrial applications. Similarly, the recourse to proteomics [68, 69], to isolate genes involved in the biosynthesis of cell wall lignin and hemicellulose in tobacco. Variations in these constituents can affect the fiber quality and cellulose availability. This could then lead to a new orientation on molecular selection research as well as genetic modifications studies to improve the quality of plant fibers.

Morphology and surface behavior of plant fibers are studied using various techniques such as XRD, FTIR, SEM, AFM, TEM and thermogravimetric analysis that helps in understanding the nature of natural fibers.

In terms of the mechanical behavior of plant fibers, important milestones have been achieved to highlight the influence of the chemical composition and structural parameters of the plant wall on their tensile properties. The microstructure of plant fibers is very complex, precisely when it comes to defining generalizable geometric and analytical models that describe it. As mentioned above, improving the mechanical properties of fibers may require the introduction of new types of fibers. And we could mention in this regard the ongoing research on spinning with solvents [70, 71], to obtain fibers of greater strength and low scattered properties. Understanding how fiber morphology affects the properties of composite materials is essential. More precisely, it is important for the selection of new fibers and for the cultivation of fibrous plants genetically selected. This would help to predict their potential for reinforcement in other materials to achieve desired properties.

Investigation of the viscoelastic properties of plant fibers has also been outlined. A variety of dynamic modulus measurement methods exists including ultrasonic wave propagation and the flexural resonance method presented here, for which normal modes of vibration are monitored. Stress relaxation tests are to be carried out to retrieve stress over time as well as the elastic modulus of the fiber material. A mathematical method for extracting the relaxation modulus from relaxation experimental data has to be proposed to this end. Proper selection of the testing vibrational mode and machine cross-head speed (during relaxation) appear important in the suggested methods in order to avoid dispersive results. The Young's modulus that is obtained from the dynamic behavior of the specimen should, therefore, reflects the frequency dependence of the material.

Acknowledgements

The authors wish to acknowledge the Director of the University Institute of Technology of the University Ngaoundéré, Prof. Mohammadou Bouba Adji, for providing research facilities within the department of mechanical engineering.

Author details

Richard Ntenga[1*], Saidjo Saidjo[1], Annie Wakata[2], Pagore Djoda[3], Martin Tango[4] and Etienne Mfoumou[5]

1 University of Ngaoundéré, Ngaoundéré, Cameroon

2 The University of Yaoundé I, Yaoundé, Cameroon

3 University of Maroua, Maroua, Cameroon

4 Acadia University, Wolfville, NS, Canada

5 Nova Scotia Community College, Dartmouth, NS, Canada

*Address all correspondence to: rntenga@gmail.com; richard.ntenga@univ-ndere.cm

References

[1] Thomas S, Paul SA, Pothan LA, Deepa B. Cellulose Fibers: Bio- and Nano-Polymer Composites. Springer-Verlag-Berlin, Heidelberg. 2011. DOI: 10.1007/978-3-642-17370-7. ISBN: 9783642173691

[2] Persson K. Micromechanical Modelling of Wood and Fibre Properties, Lund University. Department of Mechanics and Materials: Lund, Sweden; 2000

[3] Navi P, Heger F. Comportement Thermo-hydromécanique du bois: Applications Technologiques et dans les Structures. Lausanne, Suisse: Presses Polytecniques et Universitaires Romandes; 2005

[4] Ntenga R, Béakou A, Atangana Atéba J, Ayina Ohandja L. Estimation of the elastic anisotropy of sisal fibres by an inverse method. Journal of Materials Science. 2008;**43**:6206-6213. DOI: 10.1007/s10853-008-2925-2

[5] Cichocki FR, Thomason JL. Thermoelastic anisotropy of a natural fiber. Composites Science and Technology. 2002;**62**:669-678. DOI: 10.1016/S0266-3538(02)00011-8

[6] Goudenhooft C, Siniscalco D, Arnould O, Bourmaud A, Sire O, Gorshkova T, et al. Investigation of the mechanical properties of flax cell walls during plant development: The relation between performance and cell wall structure. Fibers. 2018;**6**:1-9. DOI: 10.3390/fib6010006

[7] Hulle A, Kadole P, Katkar P. *Agave Americana* leaf fibers. Fibers. 2015;**3**: 64-75. DOI: 10.3390/fib3010064

[8] Yu Y, Jiang Z, Fei B, Wang G, Wang H. An improved microtensile technique for mechanical characterization of short plant fibers: A case study on bamboo fibers. Journal of Materials Science. 2011;**46**:739-746. DOI: 10.1007/s10853-010-4806-8

[9] Réquilé S, Goudenhooft C, Bourmaud A, Le Duigou A, Baley C. Exploring the link between flexural behaviour of hemp and flax stems and fibre stiffness. Industrial Crops and Products. 2018;**113**:179-186. DOI: 10.1016/J.INDCROP.2018.01.035

[10] Duval A, Bourmaud A, Augier L, Baley C. Influence of the sampling area of the stem on the mechanical properties of hemp fibers. Materials Letters. 2011; **65**:797-800. DOI: 10.1016/j.matlet.2010.11.053

[11] Krishna KV, Kanny K. The effect of treatment on kenaf fiber using green approach and their reinforced epoxy composites. Composites. Part B, Engineering. 2016;**104**. DOI: 10.1016/j.compositesb.2016.08.010

[12] Kompella MK, Lambros J. Micromechanical characterization of cellulose fibers. Polymer Testing. 2002; **21**:523-530. DOI: 10.1016/S0142-9418(01)00119-2

[13] Pacheco JEL, Bavastri CA, Pereira JT. Viscoelastic relaxation modulus characterization using Prony series. Latin American Journal of Solids and Structures. 2015;**12**(2):420-445. DOI: 10.1590/1679-78251412

[14] Ntenga R, Saïdjo S, Beda T, Béakou A. Estimation of the effects of the cross-head speed and temperature on the mechanical strength of kenaf bast fibers using Weibull and Monte-Carlo statistics. Fibers. 2019;7:89. DOI: 10.3390/fib7100089

[15] Saïdjo S, Ntenga R, Mfoumou E, Beda T, Beakou A. Predicting the relaxation modulus for the study of the delayed behaviour of kenaf fibres in stress relaxation. Fibres & Textiles in Eastern Europe. 2021;**29**:19-25. DOI: 10.5604/01.3001.0014.7783

[16] Akhil New Normal: Overview of Sustainable Supply Chains and Preferred Fibers. Available from: https://greenstory.ca/blog/new-normal-sustainable-supply-chains-preferred-fibers/ [Accessed: November 5, 2021]

[17] Tango M, Cook K, Njagi R, Browne R, Mfoumou E, Ghosh S, et al. Developing Bio-Economy of Flax Fibers in Atlantic Canada. Darmouth, Canada: Acadia university. 2018

[18] Akubueze EU, Ezeanyanaso CS, Muniru S, Affo G, Igwe CC. Extraction & production of agro—Sack from banana (*Musa sapientum*) & plantain (*Musa Paradisiaca* L) fibres for packaging agricultural produce. International Journal of Agriculture and Crop Sciences. 2015;**9**:9-14

[19] Paramasivam SK, Panneerselvam D, Sundaram D. Extraction, characterization and enzymatic degumming of banana fiber extraction, characterization and enzymatic degumming of Banana. Journal of Natural Fibers. 2020;**00**:1-10. DOI: 10.1080/15440478.2020.1764456

[20] Ruan P, Raghavan V, Gariepy Y, Du J. Characterization of flax water retting of different durations in laboratory condition and evaluation of its fiber properties. BioResources. 2015; **10**:3553-3563. DOI: 10.15376/biores.10.2.3553-3563

[21] Townsend T. Natural Fibres and the World Economy. DNFI/FAO. 2019. Available from: dnfi.org/coir/natural-fibres-and-the-world-economy-july-2019_18043. (file accessed on 7th nov 2021)

[22] Béakou A, Ntenga R. Structure, morphology and mechanical properties of Rhectophyllum camerunense (RC) plant fiber. Part II: Computational homogenization of the anisotropic elastic properties. Computational Materials Science. Mans, France: Thesis of the University of Mans. 2011;**50**:1550-1558. DOI: 10.1016/j.commatsci.2010.12.013. Available from: https://tel.archives-ouvertes.fr/tel-00793179

[23] Mumthas ACSI, Wickramasinghe GLD, Gunasekera USW. Effect of physical, chemical and biological extraction methods on the physical behaviour of banana pseudo-stem fibres: Based on fibres extracted from five common Sri Lankan cultivars. Journal of Engineered Fibers and Fabrics. 2019;**14**:1-15. DOI: 10.1177/1558925019865697

[24] Cabrales L, Abidi N, Manciu F. Characterization of developing cotton fibers by confocal raman microscopy. Fibers. 2014;**2**:285-294. DOI: 10.3390/fib2040285

[25] Harrington JJ, Booker R, Astley RJ. Modelling the elastic properties of softwood—Part I: The cell-wall lamellae. Holz als Roh-und Werkstoff. 1995;**56**:37-41

[26] Baley C. Renforcement des polymeres par des fibres vegetales, elements cles. Revue des Composites et des Matériaux Avancés. 2007;**16**:1-8

[27] Karray M. Caractérisation des propriétés mécaniques et électriques des matériaux composites par méthode holographique numérique 3D et analyse diélectrique. Mans, France: Thesis of the University of Mans. 2012. Available from: https://tel.archives-ouvertes.fr/tel-00793179

[28] Béakou A, Ntenga R, Lepetit J, Atéba JA, Ayina LO. Physico-chemical and microstructural characterization of "Rhectophyllum camerunense" plant fiber. Composites. Part A, Applied Science and Manufacturing. 2008;**39**:67-74. DOI: 10.1016/j.compositesa.2007.09.002

[29] Huang Y, Fei B, Wei P, Zhao C. Mechanical properties of bamboo fiber cell walls during the culm development by nanoindentation. Industrial Crops and Products. 2016;**92**:102-108. DOI: 10.1016/j.indcrop.2016.07.037

[30] Mukherjee PS, Satyanarayana KG. Structure and properties of some vegetable fibres. Part 2. Pineapple fibre (Anannus Comosus). Journal of Materials Science. 1986;**21**:51-56

[31] Tze WTY, Wang S, Rials TG, Pharr GM, Kelley SS. Nanoindentation of wood cell walls: Continuous stiffness and hardness measurements. Composites. Part A, Applied Science and Manufacturing. 2007;**38**:945-953. DOI: 10.1016/j.compositesa.2006.06.018

[32] ASTM International ASTM D3379-89 Standard Test Method for Tensile Strength and Young's Modulus for High Modulus Single Filament Fibers. USA: ASTM. 1989

[33] Charlet K, Gaillard-Martinie B, Béakou A. Comportement mécanique et modélisation numérique de la lamelle mitoyenne de lin. Annales de Chimie—Science des Matériaux. 2012;**37**: 341-350. DOI: 10.3166/RCMA.22.341-350

[34] Xu P, Liu H. Models of microfibril elastic modulus parallel to the cell axis (Wood Science and Technology (2004) 38 (363-374)). Wood Science and Technology. 2004;**38**:577. DOI: 10.1007/s00226-004-0251-7

[35] Ho THI, T.N. Etude de l'influence de la température et de l'humidité sur les propriétés mécaniques en traction des fibres de chanvre et de coco. Université du Québec; 2008. DOI: 10.1016/j.cell.2009.01.043. ISBN:9781493918621. ISSN:10974172. PMID: 20056882

[36] Siqueira G, Bras J, Dufresne A. Cellulosic bionanocomposites: A review of preparation, properties and applications. Polymers (Basel). 2010;**2**:728-765. DOI: 10.3390/polym2040728

[37] Nenonene AY. Elaboration et caractérisation mécanique de panneaux de particules de tige de kénaf et de bioadhésifs à base de colle d'os, de tannin ou de mucilage. 2009; pp. 217. Available from: http://oatao.univ-toulouse.fr/7834/1/nenonene

[38] Pyszinski W, Hejnowicz Z. Is the orientation of the fibrilar helix in the main layer of the cell walls constant or variable wthin the tree. Acta Societatis Botanicorum Poloniae. 1972;**41**:27-38

[39] Huang C, Kutscha NP, Leaf GJ, Megraw RA. Microfibril Angle in Wood. New York, Basel: Buterfield; 1998. pp. 177-205

[40] Herman M, Dutilleul P, Avella-Shaw T. Growth rate effects on intra ring and inter-ring trajectories of microfibril angle in Norway spruce (Picea Abies). IAWA Journal. 1999;**20**:3-21

[41] Lichtenegger H, Reiterer A, Stanzl Tschegg SE, Fratzel P. Variation of cellulose microfibril angles in soft woods and hard woods: A possible strategy of mechanical optimization. Journal of Structural Biology. 1999;**128**:257-269

[42] Jang HF. Measurement of fibril angle in wood fibres with polarisation confocal microscopy. Journal of Pulp and Paper Science. 1998;**24**:224-230

[43] El Hadji BL. Nouveaux matériaux composites thermoformables à base de fibres de cellulose. Institut National Polytechnique de Grenoble; 2008. Available from: https://tel.archives-ouvertes.fr/tel-00268828

[44] Sedan D. Etude des interactions physico-chimiques aux interfaces fibres de chanvre/ciment. Influence sur les propriétés mécaniques du composite. Université de Limoge; 2007. Available from: http://aurore.unilim.fr/ori-oai-search/notice/view/unilim-ori-23351

[45] ASTM B923-02 Standard Test Method for Metal Powder Skeletal Density by Helium or Nitrogen Pycnometry. USA: ASTM. 2008

[46] Amiri A, Triplett Z, Moreira A, Brezinka N, Alcock M, Ulven CA. Standard density measurement method development for flax fiber. Industrial Crops and Products. 2017;**96**:196-202. DOI: 10.1016/j.indcrop.2016.11.060

[47] Le Gall M, Davies P, Martin N, Baley C. Recommended flax fibre density values for composite property predictions. Industrial Crops and Products. 2018;**114**:52-58. DOI: 10.1016/j.indcrop.2018.01.065

[48] Peças P, Carvalho H, Salman H, Leite M. Natural fibre composites and their applications: A review. Fibers. 2018;**2**:1-20. DOI: 10.3390/jcs2040066

[49] Gindl W, Schöberl T. The significance of the elastic modulus of wood cell walls obtained from nanoindentation measurements. Composites. Part A, Applied Science and Manufacturing. 2004;**35**:1345-1349. DOI: 10.1016/j.compositesa.2004.04.002

[50] Bourmaud A, Keryvin V, Charleux L. Approche du comportement viscoélastique d ' une fibre de lin par analyse du recouvrement d ' empreintes de nanoindentation [Investigation of the viscoelastic behavior of a flax fiber by analysing the nanoindentation imprint recovery]. 2011. pp. 1-8. Availabel from: https://hal.archives-ouvertes.fr/hal-00597508

[51] Cisse O. Characterization of the hygro-mechanical behavior of elementary bast fibers from hemp [PhD thesis]. University of Franche-Comté; 2014

[52] Scida D, Bourmaud A, Baley C. Influence of the scattering of flax fibres properties on flax/epoxy woven ply stiffness. Materials and Design. Courbevoie, France: AFM, Maison de la Mécanique. 2017;**122**:136-145. DOI: 10.1016/j.matdes.2017.02.094

[53] Bourmaud A, Baley C. Rigidity analysis of polypropylene/vegetal fibre composites after recycling. Polymer Degradation and Stability. 2009;**94**: 297-305. DOI: 10.1016/j. polymdegradstab. 2008.12.010

[54] Kong X, Cai CS, Hu J. The state-of-the-art on framework of vibration-based structural damage identification for decision making. Applied Sciences. 2017;7:497. DOI: 10.3390/app7050497

[55] ASTM International. ASTM C1548-02 Standard Test Method for Dynamic Young's Modulus, Shear Modulus, and Poisson's Ratio of Refractory Materials by Impulse Excitation of Vibration. USA: ASTM. 2020

[56] Mfoumou E, Hedberg C. Static versus low frequency dynamic elastic modulus measurement of thin films. Electronic Journal "Technical Acoustics". 2006;**17**:1-16

[57] Vahedi N, Wu C, Vassilopoulos AP, Keller T. Thermomechanical characterization of a balsa-wood-veneer structural sandwich core material at elevated temperatures. Construction and Building Materials. 2020;**230**:117037. DOI: 10.1016/J.CONBUILDMAT.2019.117037

[58] Ornaghi HL, Almeida JHS, Monticeli FM, Neves RM. Stress relaxation, creep, and recovery of carbon fiber non-crimp fabric composites. Composites Part C: Open Access. 2020;**3**:100051. DOI: 10.1016/J.JCOMC.2020.100051

[59] Benvenuti E, Orlando N, Gebhardt C, Kaliske M. An orthotropic multi-surface damage-plasticity FE-formulation for wood: Part I—Constitutive model. Computers and Structures. 2020;**240**:106350. DOI: 10.1016/J.COMPSTRUC.2020.106350

[60] Malloy RA. Prototyping and experimental stress analysis. In: Plastic Part Design for Injection Molding—An Introduction. Munich: Prototyping and Experimental Stress Analysis; 1994. pp. 285-339. Available from: http://www.hanser-elibrary.com/doi/book/10.3139/9783446433748

[61] Foss SD. A method of exponential curve fitting by numerical integration. Biometrics. 1970;**26**:815. DOI: 10.2307/2528726

[62] Xu Q, Engquist B. A mathematical model for fitting and predicting relaxation modulus and simulating viscoelastic responses. Proceedings of the Royal Society A: Mathematical, Physical and Engineering Sciences. Proc. R. Soc. A 474: 20170540. 2018;**474**. DOI: 10.1098/RSPA.2017.0540

[63] Kazemirad S, Bernard S, Hybois S, Tang A, Cloutier G. Ultrasound shear wave viscoelastography: Model-independent quantification of the complex shear modulus. IEEE Transactions on Ultrasonics, Ferroelectrics, and Frequency Control. 2016;**63**:1399-1408. DOI: 10.1109/TUFFC.2016.2583785

[64] Ntenga R, Béakou A, Atangana Atéba J, Ayina Ohandja L. Estimation of the elastic anisotropy of sisal fibres by an inverse method. Journal of Materials Science. 2008;**43**:67–74. DOI: 10.1007/s10853-008-2925-2

[65] Furtado SCR, Araujo A, Silva A. Inverse characterization of vegetable fibre-reinforced composites exposed to environmental degradation. Compos. Struct. 2018;**189**:529-544. DOI: 10.1016/J.COMPSTRUCT.2018.01.092

[66] Ilczyszyn F, Cherouat A, Montay G. Nouvelle approche pour la caractérisation mécanique des fibres naturelles. In: Proceedings of the 20ème Congrès Français de Mécanique. France: AFM, Maison de la Mécanique, Courbevoie. 2011. pp. 1-6. Available from: http://hdl.handle.net/2024/46776

[67] Hashin Z. On the elastic behaviour of fibre-reinforced materials of arbitrary transverse phase geometry. Journal of the Mechanics and Physics of Solids. 1965;**13**:119-134

[68] Das PP, Lin Q, Wong SM. Comparative proteomics of tobacco mosaic virus-infected Nicotiana tabacum plants identified major host proteins involved in photosystems and plant defence. Journal of Proteomics. 2019;**194**:191-199. DOI: 10.1016/J.JPROT.2018.11.018

[69] Tan BC, Lim YS, Lau SE. Proteomics in commercial crops: An overview. Journal of Proteomics. 2017;**169**:176-188. DOI: 10.1016/j.jprot.2017.05.018

[70] Klemm D, Heublein B, Fink HP, Bohn A. Cellulose: Fascinating biopolymer and sustainable raw material. Angewandte Chemie International Edition. 2005;**44**: 3358-3393. DOI: 10.1002/ANIE.200460587

[71] Heinze T. Cellulose: Structure and properties. Advances in Polymer Science. 2015;**271**:1-52. DOI: 10.1007/12_2015_319

Section 4

Recycling of Natural Fibers

Chapter 8

Recycling of Tropical Natural Fibers in Building Materials

Huyen Bui, Mazhar Hussain and Daniel Levacher

Abstract

This chapter deals with the reuse and recycling of tropical natural fibers and their potential valorization in construction materials in a context of circular economy and sustainable development. These fibers constitute large volumes of fibrous waste resulting from various agro-food industries worldwide. Depending on their intrinsic qualities and properties (physical and mechanical), they can be reused as reinforcing material in cementitious mixes (hardened mixes) or mixes with soils (raw mixes) and molded into a prismatic brick-like shape. These bricks constitute construction materials that have mechanical and other characteristics. A whole methodology specific to the development of such materials and their optimization is presented. It includes the preparation, characterization, cutting of the fibers to the desired length and the making of the mixtures. This is followed by optimization (fibers distribution), control and quality of the fiber-reinforced material. Two examples illustrate the application of this methodology: a reinforced mortar based on coconut fibers (hardened cementitious mix) and a green brick based on sediment and oil palm fibers.

Keywords: natural fibers, recycling, sustainability, waste management, building materials

1. Introduction

Natural fibers, whether of plant, animal or mineral origin, are widely available throughout the world. The diversity and abundance of plant fibers make them a highly renewable resource. And while some plant resources need to be preserved to ensure a sustainable global food supply, a very large quantity of plant fibers remains available. This concerns invasive plants that disrupt natural ecosystems and threaten biodiversity in the long term, as well as waste fibers. Waste fibers are mostly agricultural by-products or residues from industrial manufacturing processes. They constitute a renewable resource that amounts to millions of tons per year, of which only a small part is presently recovered.

The recycling of waste fibers is a part of an environmental strategy for sustainable waste management and implements the three Rs – reduce, reuse and recycle. It aims to reduce waste, preserve natural resources, save space for disposal and/or landfill and prevent the burning and incineration commonly applied to these waste fibers thus limiting CO_2 emissions. In many cases, the voluntary incineration of waste fibers results in the production of ash which, due to its chemical properties, can be used as a binding material as an addition to the components of cement or as a substitute for it [1]. However, the ecological impact of this disposal process is quite negative. It has been observed that for some fibers, waste or not, it is possible to design innovative products with high added value. However, their recovery has a cost and requires energy and the use of other raw materials: bio-based composite materials for the automobile and other modes of transport, furniture, packaging, nanofibers, building materials [2]. However, it is the most basic building materials, i.e., bricks, blocks and tiles, that appear to be best suited to recycle natural fibers, whether short or long, intact or crushed. The incorporation of waste fibers in the manufacturing of these basic materials has little effect on the production process while improving some of the properties and eliminating the waste without additional greenhouse gas emissions. Furthermore, if the brick is unfired, as is the case with fiber-reinforced bricks or blocks made from cementitious products, it is important to minimize the use of ordinary Portland cement with a high clinker content, i.e., OPC CEM I cement. The use of cement made from industrial by-products up to low-carbon binders allows to limit or even drop the carbon footprint. And for these bricks and binders, the economic cost can be reduced by eliminating one or more waste products, fibers and industrial byproducts. This is how the concept of co-valorization was developed [3, 4], which is both economic and ecological: eliminating waste, saving natural resources and limiting the carbon footprint. The crude brick reinforced with waste fibers makes it a perfectly ecological construction element without firing and without the use of binders. It is based on the principle of eco-valorization, which is founded on the integration of the circular economy, sustainable development, the conservation and the renewability of natural resources, and ultimately the limitation of greenhouse gas emissions. This is illustrated in **Figure 1**. Eco-valorization is intended to be more environmentally friendly. The soft material of these crude bricks that bind the fibers most often comes from clay-loam soils, but the introduction of waste soils is preferable, such as sediments or dredged sludge.

The waste fibers that can be recycled into eco-friendly building materials are numerous and diversified. Their quantity is closely linked to the world production of agricultural plants. Some of the fibers are consumed almost entirely by livestock (food such as straw and flour) and industry (textiles such as flax), but the rest are considered as waste, such as palm oil or coconut fibers. In the last decade, there has

Figure 1.
Circular economy, eco-valorization, sediment and fiber waste, earth reinforced bricks [5].

been a disproportionate growth in the agro-industry, which has resulted in an expansion of crops and consequently the production of waste fibers, as shown in **Table 1** for oil palm fibers. In the same **Table 1**, it can be seen that natural fibers of tropical origin alone constitute a huge potential of fibrous materials for recycling.

The recycling of waste fibers into building materials implies an industrial process to use a sufficient quantity of fibers over time i.e., renewability of the resource, which is why natural tropical fibers are of great interest. To ensure and maintain a quality manufacturing process, a methodology must be followed. It can be simply illustrated as in **Figure 2**.

This chapter demonstrates the importance of natural fibers in renewable and environmentally friendly building materials and also, the availability of fibers (introduction). Section 2 discusses the variability of shape, i.e., aggregates or fibers, structure (internal and external), intrinsic properties and applications of natural fibers. Section 3 gives background information on the process of fiber extraction, processing and methodologies for determining the main characteristics of fibers useful for use in building materials. Two applications are thoroughly described, one for a fiber-reinforced mortar (Section 4) and the other for fiber-reinforced raw earth, a truly ecological material (Section 5). The chapter concludes with a discussion on the advantages and shortcomings of tropical natural fibers as reinforcement materials.

In detail and accordance with **Figure 2**, the identification of the resource is necessary before any action of recycling waste fibers, this is the focus of Section 2 of the chapter "natural fibers and tropical fibers". This identification must be more complete with the knowledge of the properties of the waste fiber and its intrinsic characteristics useful for its future material recovery. These characteristics are obtained from specific tests carried out on these fibers and in particular, on natural tropical fibers such as oil palm and coconut fibers, see Section 3. The material recovery considered for these tropical waste fibers concerns the production of eco-materials for applications that are primarily local, i.e.,

Year	Tropical natural fibers				Other natural fibers*		
	Banana	Coconut	Palm oil fruit	Sugar cane	Flax	Hemp	Wheat
2009	103	61	216	1673	0.661	0.485	683
2019	116	62.4	411	1950	1.085	0.174	765

**subtropical and temperate climate zones.*

Table 1.
Production of main agricultural products as a potential natural fiber resource in Mt. [6].

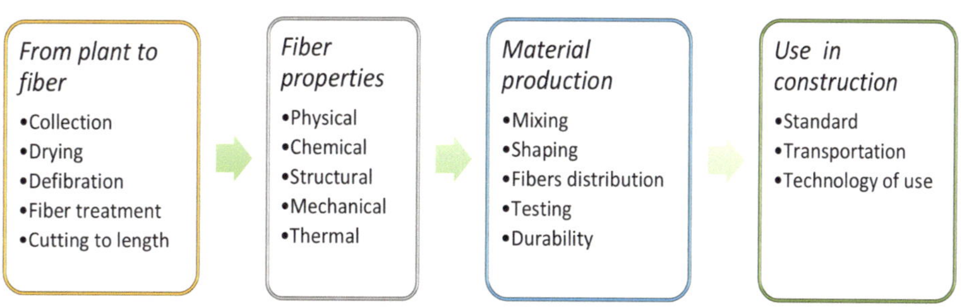

Figure 2.
A certain methodology for recycling waste natural fibers in building materials.

close to the sources of waste fiber collected. A case study of a mortar based on coconut fibers is reported in Section 4. In particular, this mortar uses calcium sulfoaluminate cement with a 37% smaller carbon footprint than Portland cement. The development of mud bricks based on oil palm waste fibers incorporated into dredged river sediment is an example of a possible eco-valorization in Section 5. These two studies demonstrate that the recycling of waste fibers into building components is potentially possible and beneficial for sustainable development.

2. Natural fibers and tropical fibers

2.1 The use of natural and tropical fibers in building materials

The use of these fibers has been temporarily set aside in preference to so-called modern cementitious materials (concretes, mortars, plasters). The usage of fibers, due to their renewability and their eco-friendly nature, is raising new interest among builders because they have interesting properties for construction. A new category of fiber-based construction materials is emerging in the field of construction and restoration: these are bio-sourced materials. Natural fibers are diverse and available in

Type of fibers	Density (g/cm^3)	Absorption coefficient (%)	Elasticity modulus (GPa)	Tensile strength (MPa)	References
Temperate climate and subtropical fibers					
Bamboo	0.45–1.3	40–145	2.82–54	39.5–1000	[7–9]
Cotton	1.21–1.6	—	1.1–13	265–800	[10]
Flax	1.19–1.55	63–330	4.4–110	93–2000	[2, 11–13]
Hemp	1.07–1.50	85–415	10–90	159–1264	[2, 10, 14]
Jute	1.23–1.50	84–281	2.5–78	300–800	[2, 10, 15]
Palm date	0.92	133–140	1.9–85	58–678	[16]
Ramie	1–1.58	—	23–128	400–1620	[10]
Reed	0.54–0.94	—	35.9	112–503	[17]
Rice straw	0.86–1.11	52–84	3.3–26.3	435–450	[18]
Sisal	1.2–1.50	110–230	1.46–38	80–1002.3	[2, 10, 14]
Wheat straw	1.14–2.05	96–320	1.4–4.8	3.45–140	[19]
Tropical fibers					
Banana spine	0.31–1.36	134–282	3–32	49.3–914	[2, 10, 17]
Coconut-coir	0.67–1.52	63–180	0.628–28	15–593	[2, 14, 20]
Palm oil*	0.1–1.55	54–120	0.5–25	147–400	[2, 5, 21]
Sugar cane	0.31–1.31	102–219	15–27.1	20–290.5	[2, 10]

**Palm oil flower and fruit are all considered.*

Table 2.
A review of physical and mechanical properties of natural fibers.

large quantities, mainly from the residues of large-scale agricultural production. The most commonly used natural fibers in building materials are, whether tropical or not, are straw (wheat, rice), flax, hemp, reed, sugar cane, jute, sisal, coconut and bamboo, as listed in **Table 2**.

The natural fibers considered in this chapter come from plants and trees and are therefore of plant origin. They are essentially so-called cultivated plants and trees, i.e., they are a renewable resource whatever their periodicity, annual for plants and several years for trees. These natural fibers are composed of cellulose, hemicellulose, lignin and pectin and impurities. While cellulose is the highest constituent in mass fraction for some plant fibers, it is much lower for woody plants where the lignin content increases or even exceeds the cellulose content (coconut, wood). The stem of the plant provides the main part of the plant fibers, whereas the leaves, fruits, seeds, bark and inter-fiber impurities are considered as plant aggregates.

Plant fibers are widely used as a building material. Over the centuries, long, flexible fibers have been used in their raw state as roofing material for rural habitats. These include wheat straw, rice straw, rushes and reeds, bamboo ... and not included in the use of plant leaves such as palm leaves, banana leaves ... But it is by combining soils and short fibers (a few centimeters) that building materials have been developed at a regional scale. They are made up of raw short fibers mixed with soils that are sometimes clayey and silty with the presence of coarse grains. These are filling materials such as cob and wooden beams; raw earth materials for the construction of load-bearing walls such as cob and prefabricated materials in the form of raw earth bricks such as adobe, see **Figure 3**. In recent decades, these plant fibers have given rise to investigations leading to more efficient construction materials, especially in terms of thermal insulation (plant fiber panels and blocks) and strength by reinforcing fibers in composite materials with a soil or cement matrix. **Figure 4** shows these different materials, noting that rice husk is a plant aggregate.

Figure 3.
Raw plant fibers in building materials: a-roofing, b-cob and wood, c-cob wall and d-adobe bricks.

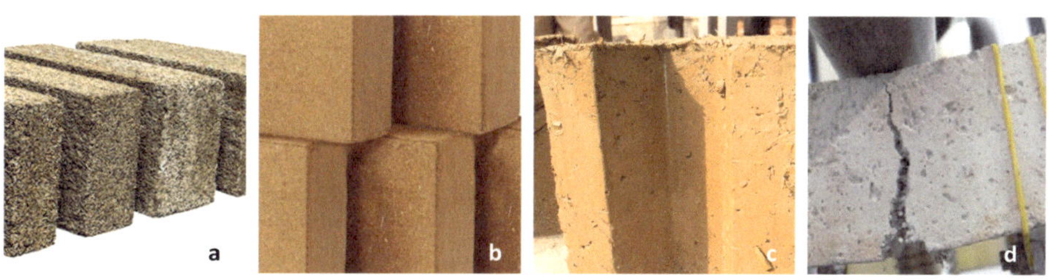

Figure 4.
Ready-to-use fiber-based insulating blocks (a-lime and hemp shiv and b-compressed coconut fibers), rice husk-based raw earth (c-rice husk) and composite mortar undergoing flexural testing (d-jute fiber [22]).

Fiber-based materials are now available in various types and shapes for construction. They are natural fibers alone, matrices of ready-to-use materials (cob), so-called efficient materials depending on the properties developed (bricks, panels, blocks), as illustrated in **Figures 3** and **4**.

Generally, plant fibers have intrinsic properties such as those related to their internal structure which gives them a high absorption capacity and hygroscopic properties. These properties are sometimes in conflict with the desired performance of the composite material being made, especially the strong performance. To achieve this, the fibers undergo a more or less chemical treatment to make them hydrophobic and improve their adhesion properties. Indeed, the external structure of the fibers plays a role in the adhesion of the fibers to the binding matrix (soils, hydraulic and similar binders, geopolymers, biopolymers, etc.). This treatment can take place at the time of the defibration process, i.e., the shaping of the fibers. In certain cases, it would delay the degradability of the fibers, and thus, improving the durability of the material.

2.2 The use of natural and tropical fibers in building materials

The selection and performance of building elements from among bio-based materials depend on the intrinsic characteristics of the fibers incorporated and the matrix containing them. These properties are thermal, acoustic, mechanical and hygroscopic. The hygroscopic character is related to both the fibers and the binding material (cob). The microstructure and biochemical composition of the fibers affect their properties as well as the treatment applied to them before their incorporation (destruction of the structure), see **Figures 5** and **6**. These fibers as mechanical reinforcement (density, length, interfacial adhesion) improve strength and performance of building materials.

Plant fibers have interesting physical properties for building materials. Due to the structure observed in **Figure 5**, these fibers have a relatively low specific density compared to metal reinforcement fibers. This is an advantage for their use, as they can produce lightweight composite materials. The bulk density is difficult to estimate due to the nature of the fiber itself as well as the geometry of the fibers, i.e., diametral dimension, cross-sectional shape. This difficulty also affects the determination of mechanical properties. The interest in plant fibers comes from their good mechanical properties, in particular a very high tensile uniaxial strength. As noted above, the determination of the ultimate strength of a fiber depends on its geometry, morphology, test operating mode (free length of the fiber, installation, loading rate), the variety of fiber plant and the unit character of the fiber (extraction mode).

The behavior of the fibers in uniaxial tension can differ depending on the fiber structure as shown in **Figure 6** where a linear behavior is observed for treated and untreated coir fibers and an elasto-visco-plastic behavior for flax fiber. The determination of the deformation modulus in the case of **Figure 6b** is problematic. Depending on the behavior, the moduli of deformation may correspond to either the initial or final slope or a linear fit over the whole curve. Both the modulus of elasticity and the ultimate tensile stress is expressed as a range of data for a fiber type due to the natural variability of fibers.

Table 2 gives an overview of these data ranges for density, absorption coefficient, modulus of elasticity in tension and tensile stress at failure for different natural and tropical fibers. Fibers in building materials are widely used as the main component either as a protective covering (braided, woven fibers) or as an insulating material

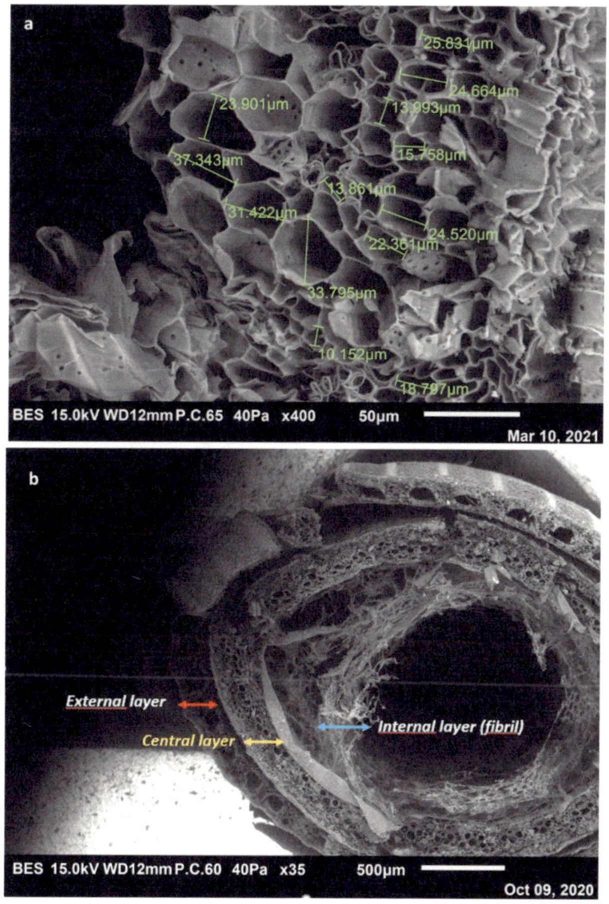

Figure 5.
SEM images of a flax straw (a) and a reed fiber (b).

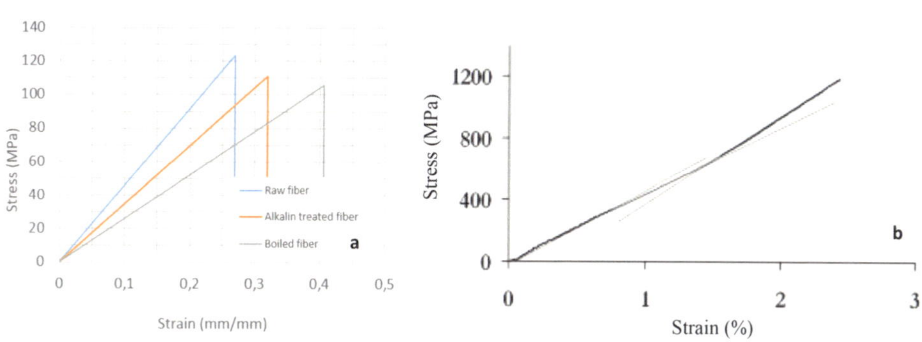

Figure 6.
Typical stress-strain relationships for (a) coconut raw and treated fibers (length 10 mm, speed rate test 0.5 mm/min, [20]) and for (b) flax fiber (length free of fiber 10.9 mm, speed test 1 mm/min [11]).

(pressed, heat-bonded, impregnated fibers). But they are also used in smaller quantities in the composition of building materials as reinforcing material. They are then randomly mixed into a binding matrix (soils, mortars or concretes).

3. Characterization and properties of tropical natural fibers

3.1 Useful properties of natural fibers in building materials

This section includes the discussion on the physical and mechanical characteristics of natural fibers which are important for their recycling in construction materials. Water absorption of fibers has a significant influence on the strength of building composites as swelling and shrinkage of fibers with their interaction with water introduce cracks in building composites. There is also a competition between the water required for hydration in the case of cementitious matrices reinforced with natural fibers and the water absorption-desorption of the incorporated fibers. Natural fibers have a low density which makes them suitable additives in lightweight building composites. Geometrical characteristics of fibers include length, diameter, surface roughness and cross-section. Increasing fibers length has a positive impact on the tensile strength of building composites however their compressive strength decreases with long fibers. Diameter of technical fiber increases with increasing elementary fibers which increases the tensile strength of technical fibers. Alignment of elementary fibers at different microfibers angles is also important for tensile strength of fibers. Higher tensile strength of fibers increases the tensile strength and toughness of composite materials. Surface roughness of fibers is essential for bonding between matrix and fibers in composites. The pull-out strength of the fibers shows how the interfacial bonding of the fibers with the matrix behaves. Failure and sliding of fibers at different loads and depths change the resistance and failure mechanism of building composites. Life and performance of fibers decrease with time. This is rapid in an alkaline environment associated with concrete structures. Treatment of fibers improves the shelf life and resistance.

Natural plant fibers are used or recycled with or without treatment. Whatever their future use, they are subject to biochemical and physical, morphological and mechanical characterisations. The biochemical characterization concerns the evaluation of the quantities of cellulose, hemicellulose, pectin, lignin, impurities and water. The physical parameters investigated include density, dimensional aspect, geometry, microfibrillar angle and water absorption capacity. The analysis of the morphology of the fibers focuses on the microstructure, the dimensional variations (diameter, length and section). The mechanical parameters sought are most often related to the traction of the fibers (isolated or anchored in a matrix). In this section, only a few procedures (treatment) or characterization tests intended for the use or recycling of fibers in construction materials are mentioned.

3.2 Treatment of fibers

The treatment of natural fibers is part of the preparation process. It takes place at the level of defibrillation (fiber separation), removal of impurities (fiber cleaning), reduction of absorption capacity (fiber hydrophobicity) and improvement of fiber adhesion to the matrix of a composite material (surface roughness and fiber bonding). Immersion in an alkaline solution (NaOH) reveals well the cleaning of the fibers which can be observed on a bamboo fiber as well as the change in roughness for a coir fiber as shown in **Figure 7**. Fiber treatments with water, boiling water, water with organic solvents or acidic agents such as ethylene diamine tetra acetic acid (EDTA) are the most eco-responsible treatments [8, 10, 14, 20]. Physical treatments are to a lesser degree environmentally acceptable but energy consuming (autoclave treatment,

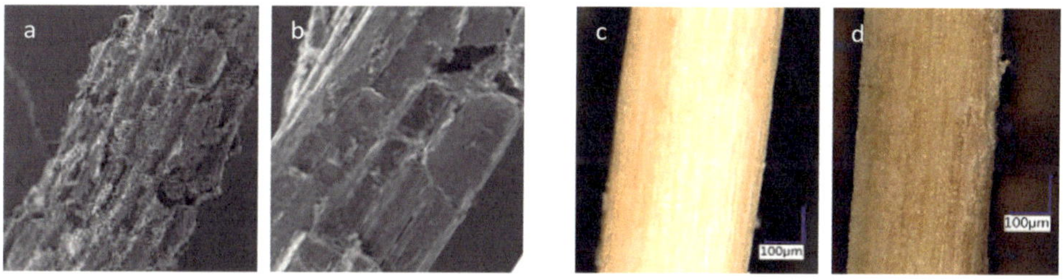

Figure 7.
Aspect of treated fibers: (a) raw bamboo fiber and (b) bamboo fiber treated for 3 days in 1% NaOH solution [23]—(c) raw coir fiber and (d) coir fiber immersed for 30 minutes in a 5% NaOH solution [20].

steam explosion, plasmas, Corona technique ...). But chemical treatments based on Na_2S, Na_2CO_3, NaOH solutions pose the problem of wastewater treatment. Some other treatments include impregnation, coupling, grafting, acetylation, benzoylation, esterification, etherification, liming .... In the most applications, alkaline solutions (**Figure 7**) and coupling techniques are the most applied. A promising technique based on enzymatic transformations would allow a more ecological treatment. These biological treatments are naturally focused on the development of biocomposites.

The benefits of these treatments are hydrophobicity, modification of the external surface of the fibers for better adhesion and improvement of the durability. The geometry of the fibers changes (decrease of diameter), they lose their flexibility but the mechanical characteristics are more or less similar.

3.3 Water absorption of fibers

Natural fibers have a very high-water absorption capacity, due to their microstructure. This absorbed water poses a problem in the elaboration of fiber-reinforced composite materials with a cementitious matrix (mortar, concrete) or raw fiber earth (adobe, cob). For the former, the water in a mixture must participate in the hydration and in the raw earth, the percentage of water is necessary for the kneading and the preparation for optimal compaction. But in the drying phase, whatever the type of material, the water contained in the fibers will be extracted and the fibers will shrink with a risk of loss of adhesion between the fiber and the matrix as shown in **Figure 8**. This amount of water absorbed must be known when making the material-fiber mixtures and the parameter to be determined is the water absorption capacity W_a expressed in %, defined by:

$$W_a\,(\%) = \frac{Mass\ of\ saturated\ fiber - Mass\ of\ dry\ fiber}{Mass\ of\ dry\ fiber} \times 100\% \tag{1}$$

The methods of water absorption determination are not always standardized. They consist of immersion in water and then measuring the mass of the fibers as a function of time. Each time the mass of the fibers is measured as a function of a time step, it is necessary to wipe them out (**Figure 9**). Various procedures are used such as the use of filter paper, fine synthetic fabric as a bag, tweezers or tea balls or manual wiping of the fibers. However, one protocol can be recommended for bio-based materials: RILEM TC 236-BBM (immersion and then dewatering for 15 seconds at 500 rpm, by means of the centrifuge). Some ranges of W_a values are given in **Figure 9**.

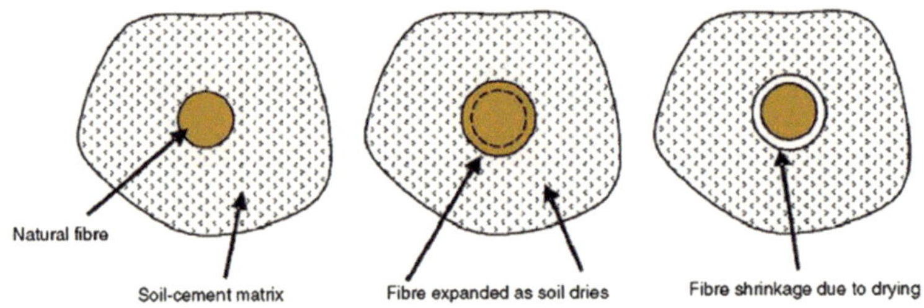

Figure 8.
Behavior of a fiber at the interface of a matrix, from the mixing phase to the curing and/or drying: illustration in the case of a cementitious material [24].

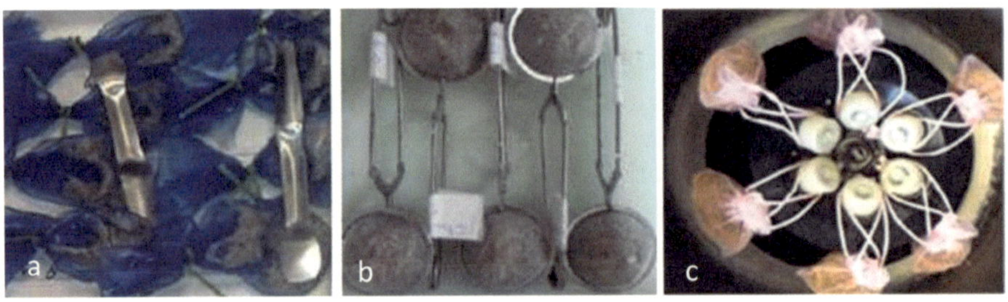

Figure 9.
Fabric bags and clips (a) for plant aggregates, tea baskets (b) for plant fibers, fiber bags (c) for centrifugal spinning and Wa ranges for different fibers [20, 25].

Alfa fiber	Banana	Coir	Date palm	Flax	Hemp	Jute	Hay	Kelp	Kenaf	Rice husk	Sisal	Straw
460-640	134-282	63-180	133-140	63-330	85-415	84-281	78-90	80	285	52-84	110-230	96-320

3.4 Specific gravity of fibers

The measurement of the absolute density or the material constituting the fiber is normally done using a helium pycnometer. But some authors still use the water pycnometer. The problem is that water, depending on the structure of the fiber, does not penetrate all the pores (underestimated density) and the absorption capacity of the fibers may overestimate the density. To avoid these phenomena, gas pycnometer (helium) is more realistic. Specific gravities differ from each type of fibers as reported in **Table 2** from the literature.

3.5 Geometry of fibers

The dimensions useful for defining the geometric parameters of a fiber are its length and diameter. The length is measured after the fibers have been cut to length by a manual (laboratory scale) or mechanical (industrial scale) process. It can be measured directly on a number of selected fibers (manual procedure) or in a more representative way, the distribution of the fibers is analyzed from a volume or a large number of fibers. This distribution gives rise to a statistical analysis (histograms, distribution law, median length...). The measurement of the diameter is more problematic because of its micro-structure (compressibility, porosity) and morphology (shape). The shape of the fibers

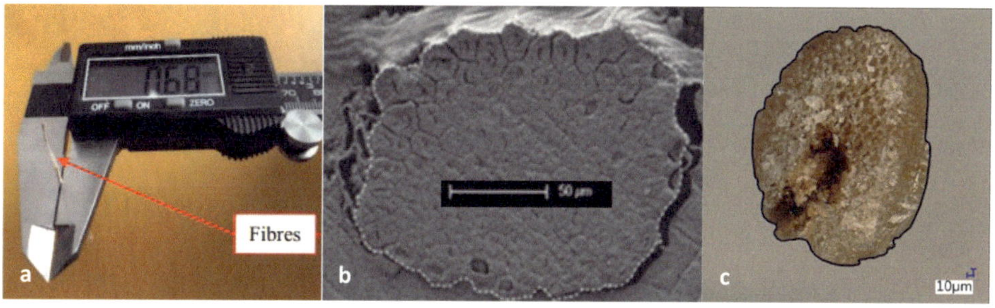

Figure 10.
Dimensions measurements: fiber diameters with digital caliper [18] (a), fiber areas with digital optical microscope (b) for a bamboo fiber [26] and a coir fiber [20] (c).

can be circular, an ellipse, an oblong shape ... depending on the type of fiber but also on the mechanical treatment: separated, crushed, shredded fiber, ... The measurement of diameters, of orthogonal axes allows to define the cross-sectional area of a fiber. But if the shape is arbitrary, the most suitable means of measurement is image analysis using a digital optical microscope associated with image processing software. From microscope images, several geometric quantities can be defined: the largest and smallest dimension (flattening coefficient), the perimeter, the area of the fiber. A more advanced exploitation allows to approach the porosity at the level of the cross-section observed. The length and the diameter of the fibers can be measured with the help of a steel rule and digital caliper respectively for diameters of the fibers as shown in **Figure 10a**. With these measurements, the fiber aspect ratio, i.e., ratio of the length to the diameter of the fiber can be determined and it is useful in the implementation of fiber composites materials.

Also, to determine the ultimate tensile stress at failure, it is necessary to know the fiber cross-section. Two methods are proposed depending on the type and shape of fibers. The first consists of taking two measurements using a caliper with a usual accuracy of ±0.01 mm that makes it possible to obtain the dimensions of the two axes of a disk or an ellipse (assumed cross-sections). The second method more representative of the shape of fibers, is essentially based on microscope image observation. Indeed, once the fiber is broken, an optical microscope is used to obtain an image of the cross-section. This image is then processed by computer-aided drawing software to determine the area of the fiber cross-section as shown in **Figures 10b** and **c**. This second method is applied in studies of the distribution and orientation of fibers within fiber-reinforced cementitious matrices or crude earth. Fiber counting in a cross-section of the composite material allows the counting of fibers in the cross-section but also the study of the observed shapes gives the orientation of fibers in the matrix. The measurement of fiber orientation by the image analysis technique requires the preparation of a material sample cross-section depending on the technique used according to Fu et al. [27]. The spatial position of a fiber can be defined by the two Euler angles θ and φ as shown in **Figures 11a** and **b**, where θ is the angle that the fiber makes with the normal direction 1 of a plane on which the fiber orientation will be observed. φ is the angle of the fiber projected in the 2–3 direction plane. θ is given by the inverse cosine of the ratio b/a (ellipse axes) and φ by the orientation of ellipse axis a to the 2-axis.

3.6 Tensile strength of fibers

One of the mechanical properties of interest is the ultimate tensile strength of the fibers. This strength value is useful in the development of composite materials. But

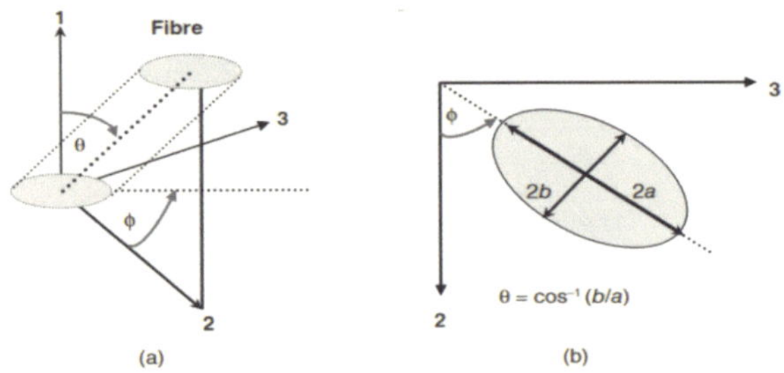

Figure 11.
An example of definitions and determination of the fiber orientation θ and ϕ angles according to Hine [28] and Fu et al. [27].

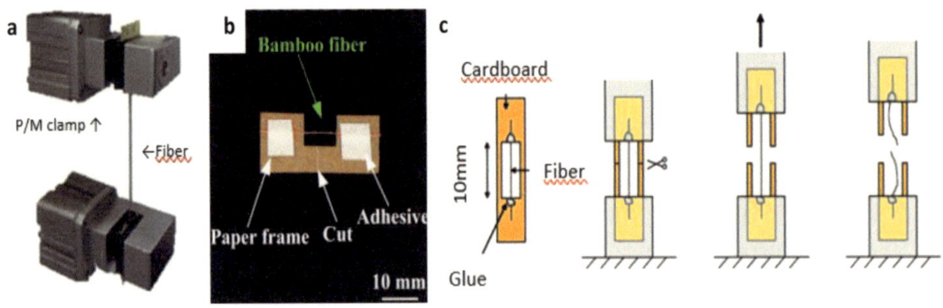

Figure 12.
Tensile strength test: possible elementary fiber installation [(a) [29], (b) [9], (c) [11]].

knowing the tensile behavior law of a free fiber (or gauge fiber) is needed for any development of numerical modeling for these materials. This behavior law is often defined by the stress-strain relationship. It is obtained from the tensile force-displacement relationship recorded during a tensile test on a fiber. To carry out the tensile tests, it is necessary to install the fiber on specific support if the fiber flexibility is limited and becomes too brittle. If the fiber is sufficiently flexible, the fiber can be clamped directly in the jaws of the testing machine. The clamping system is mechanical (M) or pneumatic (P) as shown on **Figure 12a**. But usually, for short fibers, a cardboard is used to hold the fiber before testing (**Figures 12b** and **c**). To install fiber on a cardboard frame, squares or rectangles of card stock are cut and prepared with internal dimensions depending on the free length of fiber testing. The test procedure is presented in **Figure 12c** and is as follows: installation of fiber on cardboard, clamping the cardboard on the machine, cutting the cardboard, putting the fiber under tensile loading till failure. The data recorded concerns load versus axial displacement and mainly the ultimate tensile strength as well as the maximum elongation at failure. Usually, tensile tests are carried out on different machines using different sensors. The test is performed at various constant speed rates ranging from 0.5 mm/min to 5 mm/min. Also, the machines are equipped with different more or less accurate sensors. Tests are conducted in constant room thermo-hygrometric conditions (temperature around 20–25°C). For short fibers (total length ≤ 50 mm), the free length varies from 10 to 20 mm.

Once the test is validated (failure in the part of free fiber) the stress-strain curve is analyzed and another parameter is determined: the modulus of elasticity if the fiber has an elastic or pseudo-elastic behavior. Depending on the behavior of the fiber, a linear part exists or not. It can be defined then an initial tangent modulus E_t (**Figure 13a**), or the modulus of deformation can be defined on the linear part just before the failure

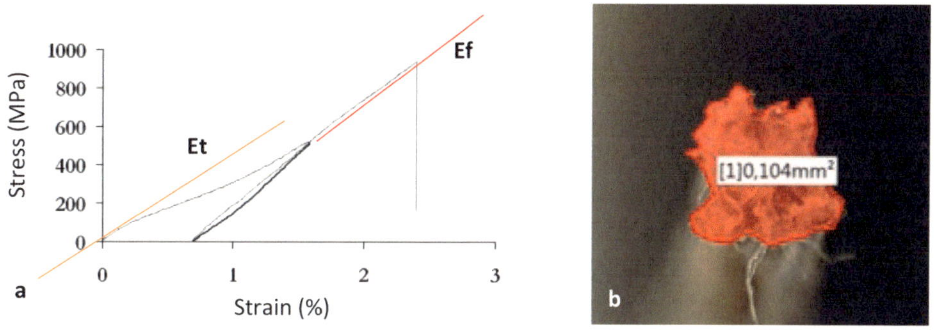

Figure 13.
(a) Flax fiber behavior under tension cycle [11], (b) determination of coconut fiber cross-section after a tensile strength test [20].

as shown in the same **Figure 13a** (modulus E_f). And this choice can be justified by the fact that a cyclic test can demonstrate elastic behavior as for a flax fiber, see **Figure 13a**. Furthermore, the determination of the ultimate stress in a fiber under traction requires the knowledge of the cross-section at the moment of rupture, although there is a constriction of the cross-section as shown in **Figure 13b**, for which the determination of the cross-section is made by using microscope image and image analysis software.

3.7 Pull-out resistance of fibers

The pull-out strength of the fibers in the matrices in which they are incorporated is another mechanical parameter necessary for the formulation of composite materials. In particular the shear stress at the fiber/matrix interface. It plays a major role in the case of short fibers [27]. It is also used to evaluate the critical fiber length. The critical fiber length (L_c) is the minimum length required to effectively strengthen and stiffen the material. It is defined by:

$$L_c = \tfrac{1}{4}\,\sigma_{tf}\ D / \tau \qquad (2)$$

where σ_{tf} is the ultimate tensile strength of the fiber, D is the fiber diameter, and τ is the interfacial shear strength at the fiber/matrix interface, see **Figure 14e**.

The critical fiber length can be estimated using the measured fiber diameter D and the values of σ_{tf} and τ issued from experimental tests or literature. The isolated (single) fiber pull out test requires a particular molding of anchored unit fibers of length L_f as shown in **Figure 14e**. The unit fibers are distributed along with a cast matrix (in the case of a cementitious or polyester resin-based material, **Figure 14d**) or crude earth (**Figures 14a** and **b**). The samples thus produced (**Figures 14c** and **d**) are submitted to a tensile test until the fiber is pulled out ($L_f < L_c$) or the fiber breaks ($L_f > L_c$). The test machines are the same as those used for the fiber tensile tests (Section 3.6).

3.8 Some properties of natural tropical fibers

3.8.1 Useful properties of natural tropical fibers in building materials

Natural fibers from the Tropics for use in building materials are relatively abundant, as the data in **Table 2** show. But among the fibers incorporated in building materials are

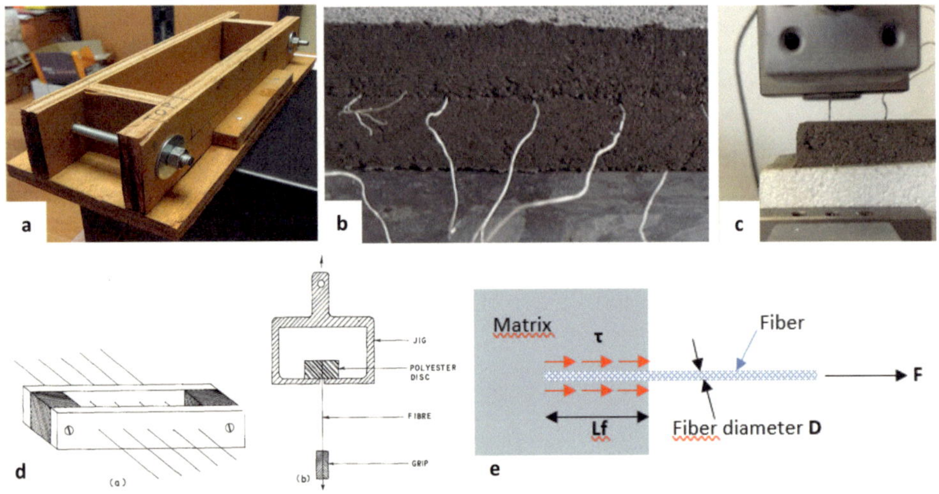

Figure 14.
(a) Crude earth specific wooden mold for pull out a test of hemp fiber, (b) crude earth sample with different hemp fibers before pull out testing, (c) pull out test of hemp fiber, (d) coir polyester composite specific mold for pull out test [30], and (e) simple mechanism of shear stress and pull-out force in the case of elementary fiber in a matrix.

coconut, sugarcane, sisal, palm fibers and to a lesser degree banana spine fiber. These are also the most widely investigated fibers in building materials at present. More recent interest has focused on the recycling of natural fibers considered as waste, such as oil palm fibers after the production of oil from the fruit. The characteristics of tropical fibers detailed below have been the focus of studies conducted by the authors. They are the most widely used fibers in building materials and in particular, the use of palm fibers constitutes an innovation in eco-friendly building materials.

3.8.2 Bio-physical properties of natural tropical fibers

Table 2 shows great variability in the data related to natural fibers, which is explained by the morphology and composition of each fiber, whether treated or not. Five types of tropical natural fibers were specifically investigated which are palm oil flower (POFl), palm oil fruit (POFr), coconut from the outer shell (Cn), sugar cane bagasse (Sc) and banana stem (Bs) fibers. These fibers are taken from the state of Tabasco in Mexico. Sugarcane fibers are separated into coarse (Scg) and fine (Scf) fibers for the study. Palm oil flower and fruit fibers are also separately considered even if some of their properties are closely related, see **Table 3**. They were not processed but extracted manually. Only the banana rachis required soaking in water and then drying of the isolated fibers.

Fibers	Density (g/cm^3)	W_a (%)	K (W/mK)	Cellulose (%)	Avg. area (mm^2)
POFL	1.37	235	0.058	48.84	0.070
POFR	1.36	258	0.055	37.36	0.027

Note: W_a = water absorption, K = thermal conductivity, and Avg. = average.

Table 3.
Some bio-physical properties of palm oil flower and fruit fibers.

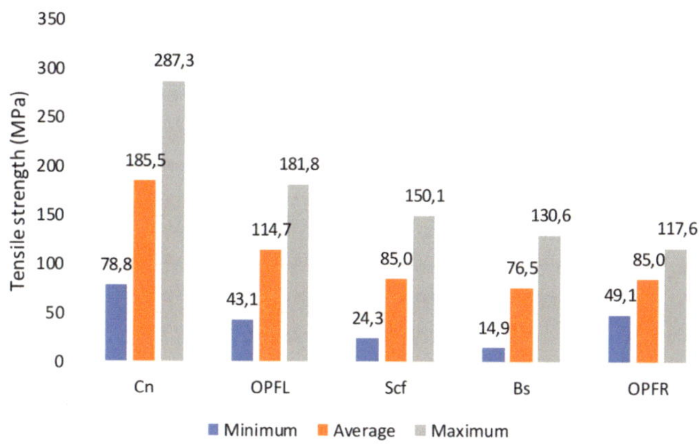

Figure 15.
Tensile strengths observed on typical tropical fibers from Mexico.

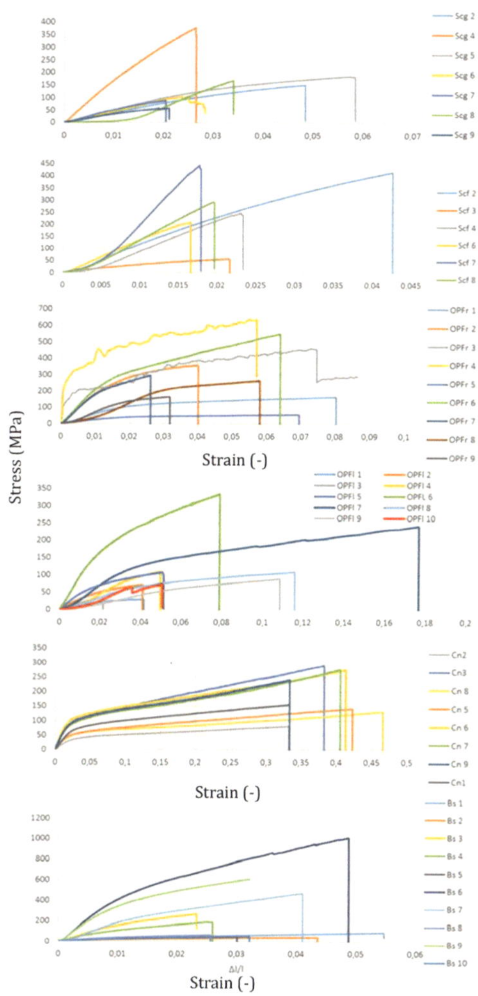

Figure 16.
Typical stress-strain relationships for different tropical fibers.

Table 3 shows the bio-physical properties of palm oil fibers such as density, water absorption, thermal conductivity, cellulose content and area. The low density of fibers makes them a suitable additive in manufacturing lightweight building composites. Natural fibers are hydrophilic materials and higher water absorption of fibers in composite leads to micro-cracks growth in composites. Natural fibers have a low thermal conductivity which has a positive influence on the thermal behavior of building materials. Higher cellulose content of fibers contributes significantly to the tensile strength of fibers. The area of fibers is important to study the mechanical characteristics of fibers such as tensile strength. Tensile strength of fibers is heavily influenced by the area of fibers as technical fibers tested in this study are consist of the number of elementary fibers.

3.8.3 Bio-physical properties of natural tropical fibers

The five types of fibers were subjected to a tensile test. The length of the fiber chosen is 20 mm (free or gauge length) for a total length of 30 mm. The constant test speed was 0.5 mm/min at a room temperature of 22 ± 2°C and relative humidity of 40–50%. A minimum of 10 fibers were tested for each series. The minimum, average and maximum tensile strength of Mexican tropical fibers is presented in **Figure 15**. The tensile and strain loading curves reveal different mechanical behaviors depending on the fiber. Pseudo-elastic, visco-elastic or elasto-plastic behavior with a strain-hardening effect can be observed in **Figure 16**.

The tensile strength of fibers discussed in **Figures 15** and **16** is one of the most important parameters for their use in bricks as fibers act as reinforcement. The higher tensile strength of fibers increases the tensile strength of building composites.

4. Application in reinforced mortars by natural fibers as cementitious material

An innovative solution to reduce the negative impact on the environment is the production of structural reinforced composites from these natural fibers. Section 4 introduced an alternative binder to improve the durability of these plant fibers in an alkaline environment of cementitious composite. CSA cement was used in coconut fiber-reinforced mortar to totally replace the traditional cement. This section is divided into 2 parts including the *mechanical properties of fiber-reinforced mortar*, and *carbonation resistance of fiber-reinforced mortar*. In each part, the comparison between unreinforced- and reinforced mortars is presented.

Among the natural fibers, coconut fiber is considered is a potential candidate for reinforcement in cement matrix due to its most ductile and energy absorbent properties compared to other plant fibers. Incorporation of natural fibers into cementitious composite could, therefore, constitute an alternative solution to waste management and contribute to the development of reinforced mortars by improving comfort performance in buildings. Besides, it is necessary to use alternative binders to improve the durability of these plant fibers into composites and reduce the negative impacts on the environment. In this section, the new formulations of mortar are proposed, in which the Portland cement is totally replaced by calcium sulfoaluminate cement (CSA cement). CSA cement, consisting of nearly 55% of calcium sulfoaluminate, could be considered as a clean, green and alternative binder due to its environmentally-friendly features [31]. Additionally, CSA cement contains a lower alkali content compared to

PC, i.e., the pH ranges of CSA cement and PC are 10–11 and 12–13, respectively. As a consequence, the lower pH value of CSA cement can also lead to the less natural degradation rate of the fibers in the alkaline environment of the cementitious matrix.

4.1 Mechanical properties of fibers-reinforced mortars

The mechanical properties of fibers-reinforced mortars depend on various parameters such as intrinsic properties of fibers, fiber contents, fiber distribution, fiber orientation, interfacial transition zone (ITZ), i.e., fibers and cementitious matrix adhesion. The decrease in the compressive strength of mortars with increasing fiber content is observed. A part of the explanation is that the pectin, ash, and other impurities are included in the fiber component, inducing the reduction of the bond between fibers and cementitious matrix. Additionally, the higher air content and porosity, relative to the increase in fiber content, involve a decrease in compressive strength. The combination method of coconut fibers and CSA cement in mortar significantly increased flexural strength of mortar, up to approximately 17%, which meets the desired mechanical performance since fibers are used as reinforcement. However, at the higher content of fibers (≥3% by mass of cement), the flexural strength starts suffering a slight decrease due to much more fibers being in the restricted area of the brittle cementitious phase, which leads to the significant cumulative effects on the strength of the material. In addition, frictional energy losses considerably in the wake of pulling out of fibers due to the debonding at the interface, which is partly responsible for the failure.

Figure 17 shows the typical evolution graphs of the force applied as a function of the displacement at mid-span of the specimen for unreinforced and 2% fiber-reinforced mortars. To clarify the understanding of the different periods of crack initiation and propagation in bending, five particular points corresponding to five load steps are noticed for reinforced mortar. Point A is at the end of the non-linear elasticity period (so A is also at the first of the linear period). This point shows how the normal displacement evolves in the elastic period during the flexural test. Crack has not occurred in this step, although the load reaches 55% of the maximum load. In the next step, point B represents the displacement in the linear part of the curve and corresponds to the point where crack starts appearing at the load of 85% the maximum. It should be noted that the formation and development of cracks also depend on the characteristics of supporting

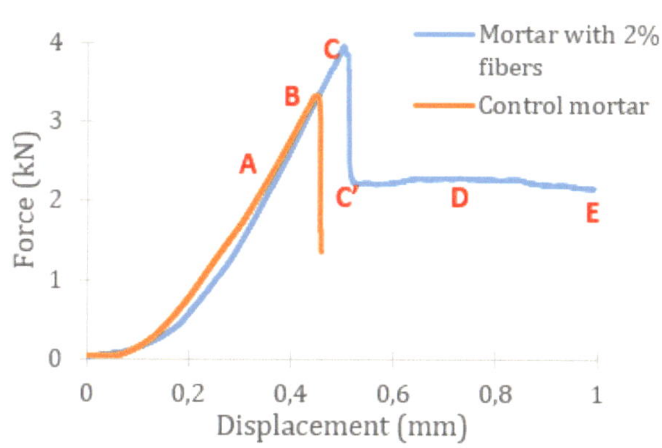

Figure 17.
Typical curves of behavior in 3 points bending of mortars [32].

(two) and loading (one) rollers of the flexural test. If one of them is capable of tilting or sliding slightly, a uniform distribution of the load over the width of the specimen is well applied. And thus, this induces the appearance of a single crack. Otherwise, multi-cracks would have occurred, and flexural behavior will be affected if all supporting rollers cannot freely rotate. Therefore, the scatter of cracks is observed on the cross-section of the sample in this case [33]. In the third step, point C corresponds to the peak of the force-displacement curve, i.e., the maximum of the flexural load. As the sample partly suddenly fails, point C′ is reached to introduce the residual force. The load reaches the maximum load, and some fibers begin pulling out from the cementitious matrix and then slip inside the mortar, as clearly shown by the drop from point C to point C′. The period from point D to point E is along the residual force step which mobilizes the shear resistance of fibers. This step describes a nearly constant load period while the bending displacement continues increasing due to the remaining fibers. The crack initiates at the base, i.e., an opposite plane to the applied load, of the sample and propagates toward the direction of loading in the wake of the appearance of the initial crack. In this stage, the contribution of fiber to preventing brittle fracture suddenly is shown clearly. Additionally, resisting fragmentation is observed as there is no spalling at the surface of the specimen due to the bridging effect of the fiber distribution. For control mortar, the bridging effect could not be observed. The sample shows a sudden drop at about 80% of the maximum applied force. The strain development of the control mortar is characterized by a non-linear elastic part followed by a nearly linear behavior before sudden failure occurs (fragile behavior). The single crack appears at the base of the samples on which it is believed to have the maximum bending moment and no shear load. The reinforced mortars show that a progressive load decrease is likely associated with a progressive rupture of the fiber-matrix interface and then limits a brittle fracture. The addition of fibers into mortar has remarkable effects on the cracking behavior of mortar. Fiber acts as a crack-arrester since the presence of fibers could contribute to preventing brittle fracture suddenly after the first crack appears. Also, the bridging effect of the fiber distribution induces a decrease in the crack width and length compared to the control sample at the same level of loading. The enhancement of toughness and preventing the development of cracks inside reinforced mortars are the most important contributions of fibers.

4.2 Carbonation resistance of fibers-reinforced mortars

In terms of durability, the usage of CSA cement with low alkali content could lead to a significant decrease in carbonation resistance owing to the lower content of CaO compared to conventional cement. Additionally, several previous studies [34, 35] also pointed out the negative effects of the Ca/Si ratio on the carbonation resistance performance. They believed that a rapid carbonation degree was acquired in consequence of the rapid decalcification of calcium silicate hydrate gel (CSH) at the higher ratio of Ca/Si. Additionally, the formation of carbonation products that result from the decomposition of ettringite, which is the principal phase of CSA cement, and contributes to boosting the carbonation depth in mortar specimens. This observation also proves that the dense microstructure formed by ettringite has negligible effects on the carbonation resistance of the CSA cement-based matrix. Besides, incorporating fibers could improve the carbonation rate due to the high air content (the fibers act as channels and entrain air), encouraging CO_2 penetration happened could be easier [36].

The various effects of carbonation on the performances of mortar were obtained. In detail, the compressive strength increased by approximately 9 and 33% for conventional cement-based mortars incorporating and no fibers, respectively. In contrast, the carbonation process could induce a slight decrease by 1–3% in compressive strength of CSA specimens with and without fibers, respectively. The pore structure of composite, which acquires significant changes after accelerated carbonation, is partly responsible for these results. It should be noted that a higher carbonation depth is found in CSA specimens. Therefore, the relationship between carbonation resistance and mechanical strength seems to be significantly dependent on the binder type used in composite [37]. Carbonation-induced strengths of mortar are various due to the cumulated effect of fibers incorporated. Mechanical behavior, hence, could not be a substantial factor in deciding the carbonation resistance of the cementitious composite.

The process of carbonation also induces a slight decrease in the thermal resistance ability of the matrix. For instance, non-carbonated zones have a strong ability to resist temperature than others in carbonation. In detail, at elevated temperature (~900°C), the carbonated area lost up to 14% of its mass. Meanwhile, the mass loss value of the non-carbonated area is below 10%. This observation is the result of $CaCO_3$ formation during the carbonation process. This compound is thermally decomposed at a temperature higher than 650°C. Otherwise, the calcium-carbonated filler, which is generated in conventional cement production, is the main phase decomposed at this temperature.

Mechanical properties of composite materials need to be assured considering the environmental vulnerability. Generally, exposure in wetting and drying cycles has strong effects on the mechanical properties of samples due to the repetition of the negative environment on the interfacial bonding between fibers and cementitious matrix. After the sample is exposed to wetting and drying cycles, compressive strength is the most critical factor in assessing the performance of composite materials [38]. The wetting and drying repetition has adverse effects on the mechanical performance of mortar, regardless of the number of fibers, and reduces both compressive and flexural strengths. Generally, losses in mechanical properties of CSA-based mortars were higher than that of PC-based mortars. However, it should be noted that the maximum compressive strength was observed after one cycle since complete hydration of cement was reached due to the addition of water during the wetting process. In the next cycles, due to the formation of crystallized hydrate products [39], more micro-cracks appeared gradually inside the mortar structure and induced a decrease in compressive strength. Both strength and deformation of mortar samples decreased at the higher level of porosity and the higher number of cycles. The loss of strength was observed when fibers were incorporated into the mortar. More pores in fine aggregate mortar appear due to adding coconut fibers, which creates a convenient environment for the deep penetration of ambient air and water. The change in mechanical strength with predicted tendency was governed by the porosity, the number of cycles and fiber content as well, i.e., the higher fiber content, the higher porosity, the higher number of wetting and drying cycles, the lower mechanical strength.

In conclusion, for natural fibers reinforced composite to become widely used construction materials, consistent and predictable results need to be obtained. To achieve these outcomes, further studies are required on these composite performances by testing and modeling, which are necessary to help the application of this material for the building materials widely. These outcomes might contribute to environmental benefits and sustainable development of the construction industries in the future.

5. Application in reinforced crude bricks by natural fibers as an eco-friendly material

The emission of greenhouse gases, global warming and environmental concerns associated with modern construction materials have forced us to look for environmentally friendly construction materials. Construction materials such as concrete, cement and fired bricks manufacturing consume a higher amount of energy and are responsible for a considerable amount of CO_2 emission. Moreover, these materials have higher thermal conductivity. On other hand, adobe bricks are environment-friendly building materials manufactured with sediments and waste natural fibers. Dredged sediments and natural fibers are renewable waste materials, easily available in most regions of the world. Disposal of waste fibers and dredged sediments have negative impacts on the environment as they are the source of air, water and land pollution. Dredged sediments and natural fibers valorisation in adobe bricks generates sustainable and green construction material as adobe bricks manufacturing does not consume energy. Adobe bricks are sun-dried and they have a low thermal conductivity which reduces the energy consumption for heating and cooling.

Adobe bricks are the cheap and oldest construction materials, used in historical buildings. Adobe bricks are manufactured with soil and natural fibers. Soil suitability for adobe bricks is observed with standards such as AFNOR and MOPT [40, 41]. Sediment's mineralogy, molding moisture content, compaction energy, clay and sand content have a significant influence on the characteristics of adobe bricks. The strength and durability limitations of adobe bricks are the obstacles to their large-scale use. The addition of natural fibers and the use of stabilization techniques reduce these drawbacks to some extent.

Natural fibers are agricultural waste generated by food industries such as sugar cane, palm oil, etc. Natural fibers act as reinforcement in adobe bricks and increase their tensile strength. The low thermal conductivity and density of natural fibers make them suitable additives for adobe bricks and composite materials. Distribution, orientation, quantity and length of fibers are some important parameters that affect the quality and performance of adobe bricks. Fibers are randomly distributed in crude bricks, and it is difficult to control their orientation. The longitudinal distribution of fibers parallel to the bricks axis increases the tensile strength of bricks while their distribution perpendicular to the bricks cross-section does not contribute to the tensile strength of bricks. The quantity of fibers has also a significant impact on the performance of composite materials. The quantity of fibers used in adobe bricks usually varies from 1–5% by mass. The length of fibers is another factor that influences the tensile and compressive strength of bricks. Length of fibers varies with the choice of fibers and available fibers cutting mechanism. The distribution of fibers in bricks is homogenous with fibers of short length. The tensile strength of bricks increases with increasing fibers length however, longer fibers harm the compressive strength of bricks. The presence of knots and weak bond in long fibers affects their performance. Therefore, the use of appropriate length fibers is suggested. The common length range of natural fibers for crude bricks varies from 2 cm to 10 cm in literature studies.

The stabilization of bricks is achieved either by compaction or by using binding materials such as lime, cement and gypsum. As the use of binders involves the cost and contributes to global CO_2 emissions. Therefore, bricks stabilization with compaction is a reasonable option. Compaction of bricks can be achieved by static loading, dynamic compaction, and tamping. The compaction of bricks removes the voids inside the bricks and improves their mechanical characteristics along with durability [42].

However, compaction of bricks with dynamic loading and tamping causes the upward movement of fibers along with water which may perturb the distribution of fibers. The durability of adobe bricks is also a challenge as the performance of fibers decreases with time due to degradation. Treatment of fibers improves their characteristics but it has environmental concerns.

Adobe bricks are manufactured by mixing sediments, fibers and water. Molding moisture content varies with the type of sediments and natural fibers. It is important to respect the water absorption coefficient of natural fibers. As water absorption of natural fibers is not instantaneous, prior fibers saturation helps to make homogenous sediments and fibers mixture.

5.1 Manufacturing of crude bricks

Dredged sediments from the Usumacinta River and palm oil flower fibers (POFL) from the Tabasco State of Mexico were investigated to use them in adobe bricks. Usumacinta River sediments (USU) are shown in **Figure 18a**. Sediment's characteristics such as granulometry, density, Atterberg limits, optimum moisture content, carbonate content, pH, XRF and organic matter were analyzed. Sediment characteristics are summarized in **Table 4**.

Palm oil flower fibers were used as reinforcement in adobe bricks. POFL fibers are obtained from palm oil empty fruit bunches which are waste material from the palm industry. Empty fruit bunches were cut with a knife mill by using a grid of 3 cm. POFL fibers are shown in **Figure 18a**. Due to the grinding of fibers in the knife mill, there is length variation for grid-3 cm long fibers. The average length of fibers is 11.54 mm.

Fiber's characteristics such as length, cross-section, tensile strength, water absorption, chemical composition and thermal conductivity are important for their use in crude bricks. Cellulose is the main component of POFL fibers which play a key role in the tensile strength of fibers. The tensile load behavior of POFL fibers is elastoplastic behavior. Characteristics of POFL fibers are shown in **Table 3**.

Figure 18.
Adobe bricks manufacturing, (a) dry sediment and fibers, (b) miniature proctor for prismatic sample, and (c) molded crude bricks.

Sediments	LL (%)	ρ_{sed} (g/cm^3)	Clay (%)	Sand (%)	pH (−)	OM (%)	MBV (g/100 g)	$CaCO_3$ (%)	Wopt (%)	SSA (m^2/g)
J3-9C	37.74	2.63	5.9	52.8	7.5	4.48	2.73	7.84	19.3	28.20

Note: LL = liquid limit, OM = organic matter, MBV = methylene blue value, and SSA = specific surface area.

Table 4.
Usumacinta sediments characteristics.

Crude bricks were manufactured with POFL fibers and Usumacinta River sediments. Manufacturing of fired bricks consists of mixing sediments and fibers, molding, compaction and drying. USU sediments were crushed, grinded and passed through a 2 mm sieve. Sediments were mixed with 0, 1, 2, 3, 4 and 5% saturated POFL fibers. 450 g sediments are recommended for prismatic bricks of size 4*4*16 cm^3. The quantity of fibers for each brick can be calculated by Eq. (3).

$$m_{\text{fibers}} = \left(m_{sed} \times \%\text{ of fibers}\right) / 100 \tag{3}$$

where m_{sed} is the mass of dry sediments, m_{fibers} the mass of fibers and % of fibers, the percentage of fibers used. The optimum moisture content of sediments was found by the Proctor test and its value is 19.3%. It was used as molding moisture content. The quantity of water for the mixture can be calculated by Eq. (4).

$$m_{\text{water}} = \left(m_{sed} \times \%\text{ of water}\right) / 100 \tag{4}$$

where m_{sed} is the mass of dry sediments, m_{water} the mass of water and % of water, the optimum water content.

USU sediments and POFL fibers were poured into a mixing bowl followed by the addition of water. Sediments and fibers were mixed with an electric mixer for 10 minutes. Sediment mixing was followed by molding. Sediment's mixture was molded into prismatic bricks of size 4*4*16 cm^3 which is commonly used for manufacturing composite materials at laboratory scale [11]. Bricks were compacted with dynamic compaction. Compaction energy used is similar to the Proctor test energy, i.e., 600 $kN.m/m^3$. The compaction of adobe bricks is shown in **Figure 18b**. After compaction bricks were unmolded and oven-dried at 40°C and sun-dried. Bricks were kept in the oven until their mass variation was below 1%. It was observed that sun-drying of bricks at room temperature (20°C ± 2°C) takes 2–3 weeks while oven drying can be achieved in 3–4 days. **Figure 18c** shows oven drying of adobe bricks.

5.2 Characteristics of bricks

Adobe bricks characteristics include tensile strength, fibers distribution, thermal conductivity, shrinkage and density, etc. Characteristics of bricks are strongly influenced by fiber content. Tensile strength and toughness of bricks increase with fiber content up to optimum moisture content. Fibers contribute to the tensile strength of bricks and transform the brittle failure into ductile failure by increasing the post-peak load-bearing capacity of bricks. The higher quantity of fibers produces fibers clusters in the matrix and reduces the bonding between fibers and sediments which decreases the tensile strength and toughness of bricks. The indirect tensile strength of bricks is determined with a three-point bending test according to ASTM standard [43]. The toughness index of bricks is also be determined with a tensile strength test. The tensile strength and toughness index of Usumacinta sediments bricks at different fiber content are shown in **Table 5**.

Table 5 shows that the tensile strength and toughness of Usumacinta sediments bricks increases with fiber content up to the optimum fiber content. Tensile strength of fiber-reinforced crude bricks, i.e., adobe bricks, reported in the literature and different standards vary from 0.04 to 2.05 MPa [44].

DCi%	DC0%	DC1%	DC2%	DC3%	DC%	DC5%
σ_t (MPa)	1.79	1.79	2.56	3.19	2.02	2.59
Toughness	1	2.58	4.18	4.42	2.83	3.89

Note: DCi% = dynamic compaction of crude brick with i (%) of fiber content, and σt = tensile strength at failure.

Table 5.
Mechanical characteristics of bricks.

Bricks	Mass (g)	Density (kg/m^3)	Shrinkage (%)	K (W/mK)	UPV (m/s)
USU	370	1524	2.25	0.23	924

Note: K = thermal conductivity, and UPV = ultrasonic pulse velocity.

Table 6.
Physical characteristics of adobe bricks.

Physical characteristics of bricks include mass, density, shrinkage, ultrasonic pulse velocity (UPV) and thermal conductivity. The addition of fibers decreases the density of bricks and makes them lightweight construction material. On the other hand, Ultrasonic pulse velocity of bricks also decreases with increasing fibers content due to the presence of voids as saturated fibers swell initially and shrink after drying the bricks which produce small cracks. Shrinkage is another important property of bricks which increases with higher molding moisture content and higher clay content of sediments.

Physical characteristics of adobe bricks made with USU sediments at optimum fiber content are summarized in **Table 6**.

Uniform distribution of fibers inside the bricks is important. The distribution of fibers inside bricks can be observed with ImageJ software [5]. Crude bricks are divided into 4 parts with 6 cross-sections. A microscopic image of each brick cross-section is analyzed with ImageJ software to find the number of fibers, their area and orientation in bricks. Upward movement of fibers in composite materials with dynamic compaction can be observed with image analysis.

This section includes a detailed analysis of the physical and mechanical characteristics of natural fibers especially tropical fibers for their recycling in construction materials such as crude bricks. Crude bricks specimens were manufactured at different fibers content and their characteristics were analyzed. The Addition of natural fibers in construction material has a positive impact on the tensile strength, density and thermal properties of these materials.

6. Conclusions

This chapter focuses on the recycling of waste natural fibers in composite materials used for construction. Fiber's recycling eliminates agriculture waste and contributes to conserving natural resources used in building materials and sustainable development.

For fibers recycling, the study of their biochemical, physical, morphological, microstructural and mechanical characteristics is essential. In addition, durability analysis of fibers helps to determine the long-term evolution of these characteristics. This chapter reviews the characteristics of natural fibers suitable for building composites such as tensile strength, water absorption, modulus of elasticity and density.

Moreover, characteristics of tropical fibers are also discussed for their reuse in earth bricks.

Analysis of natural fibers characteristics used in building materials shows the great variability in their intrinsic properties. The development of composite materials such as reinforced cementitious mortars or fiber-reinforced raw earth must consider this variability.

Natural fibers addition in building composites improves the tensile strength of these materials. Moreover, natural fibers improve the thermal and acoustic performance of composite materials.

Case studies for the recycling of natural fibers in mortar and earth bricks are also discussed. Natural fibers addition reduces the density of earth bricks and mortar. The tensile strength of earth bricks increases significantly with the addition of natural fibers. Natural fibers act as reinforcement and transform the brittle behavior of mortar and earth bricks into ductile behavior.

Acknowledgements

A part of this research work reported in this chapter has been funded by the project "From traditional uses to an integrated valorisation of sediments in the Usumacinta river basin (VAL-USES)" from the Agence Nationale de la Recherche of France (ANR-17-CE03-0012-01) and the Consejo Nacional de Ciencia y Tecnología of Mexico (FONCICYT-290792).

Author details

Huyen Bui[1,2], Mazhar Hussain[3,4] and Daniel Levacher[3,4]*

1 Unité de Chimie Environnementale et Interactions sur le Vivant, Littoral Côte d'Opale University, UR 4492, UCEIV, SFR Condorcet FR CNRS 3417, Dunkirk, France

2 Faculty of Civil Engineering, Thuyloi University, Hanoi, Vietnam

3 ComUE Normandie University, Caen Normandie University, M2C UMR 6143 CNRS, Caen, France

4 UniLaSalle, Univ. Artois, ULR 7519, Transformations and Agro-Resources, ComUE Normandie University, Mont Saint Aignan, France

*Address all correspondence to: daniel.levacher@unicaen.fr

References

[1] Maraveas C. Production of sustainable construction materials using agro-wastes. Materials. 2020;**13**(2):262. DOI: 10.3390/ma13020262

[2] Dungani R, Karina M, Subyakto SA, Hermawan D, Hadiyane A. Agricultural waste fibers towards sustainability and advanced utilization: A review. Asian Journal of Plant Sciences. 2016;**15**:42-55. DOI: 10.3923/ajps.2016.42.55

[3] Levacher D, Batagin B, Liang Y, Cherifi N. Gestion durable des sédiments: co-valorisation sédiments-fibres. In: Proceedings of Journées Nationales Génie Côtier-Génie Civil (JNGCGC2012); 12-14 juin 2012. Cherbourg, France; 2012. pp. 1025-1036. DOI: 10.5150/jngcgc.2012.113-L

[4] Pereira CL, Savastano H Jr, Payá J, Santos SF, Borrachero MV, Monzó J, et al. Use of highly reactive rice husk ash in the production of cement matrix reinforced with green coconut fiber. Industrial Crops and Products. 2013;**49**:88-96. DOI: 10.1016/j.indcrop.2013.04.038

[5] Hussain M, Levacher D, Leblanc N, Zmamou H, Djeran-Maigre I, Razakamanantsoa A. Influence of palm oil fibers length variation on mechanical properties of reinforced crude bricks. Construction Technologies and Architecture. 2022;**2022**(1):707-714

[6] FAO. Statistic of crop production in the world – Crops and livestock products (Production). Rome: Food and Agriculture Organization; 2021. Available from: https://www.fao.org/faostat/en/#data [Accessed: 2021-11-22]

[7] Rusch F, Bordignon Ceolin G, Hillig E. Morphology, density and dimensions of bamboo fibers: A bibliographical compilation. Pesquisa Agropecuária Tropical. 2019;**49**(1):9. DOI: 10.1590/1983-40632019v4955007

[8] Shah AUM, Sultan MTH, Jawaid M, Cardona F, Talib ARA. A review on the tensile properties of bamboo fiber reinforced polymer composites. BioResources. 2016;**11**(4):10654-10676. DOI: 10.15376/biores.11.4.Shah

[9] Zhang K, Wang F, Liang W, Wang Z, Duan Z, Yang B. Thermal and mechanical properties of bamboo fiber reinforced epoxy composites. Polymers. 2018;**10**(6):608. DOI: 10.3390/polym10060608

[10] Thyavihalli Girijappa YG, Mavinkere Rangappa S, Parameswaranpillai J, Siengchin S. Natural fibers as sustainable and renewable resource for development of eco-friendly composites: A comprehensive review. Frontiers in Materials. 2019;**6**:226. DOI: 10.3389/fmats.2019.00226

[11] Charlet K. Contribution à l'étude de composites unidirectionnels renforcés par des fibres de lin : relation entre la microstructure de la fibre et ses propriétés mécaniques [thesis]. France: Caen University; 2008

[12] Baley C. Analysis of the flax fibers tensile behaviour and analysis of the tensile stiffness increase. Composites: Part A. 2002;**33**:939-948. DOI: 10.1016/S1359-835X(02)00040-4

[13] Davies GC, Bruce DM. Effect of environmental relative humidity and damage on the tensile properties of flax and nettle fibres. Textile Research Journal. 1998;**68**:623-629. DOI: 10.1177/004051759806800901

[14] Shah I, Jing L, Fei ZM, Yuan YS, Farooq MU, Kanjana N. A review on chemical modification by using sodium hydroxide (NaOH) to investigate the mechanical properties of sisal, coir and hemp fiber reinforced concrete composites. Journal of Natural Fibers, 2021;1-19. DOI: 10.1080/15440478.2021.1875359

[15] Sridhar SK, Basavarajappa G, Kasturi SG, Balasubramanian N. Evaluation of jute as reinforcement in composites. Indian Journal of Textile Research. 1982;7(3):87-92. http://nopr.niscair.res.in/handle/123456789/33127

[16] Masri T. Contribution au développement des matériaux de construction à base des sous-produits du palmier dattier [thesis]. Algeria: Biskra University; 2018

[17] Tian KP, Shen C, Zhang B, Li XW, Huang JC, Chen QM. Experimental study on mechanical properties of reed stalk. IOP Conference Series: Earth and Environmental Science. 2019;**346**:012076. DOI: 10.1088/1755-1315/346/1/012076

[18] Danso H. Properties of plantain pseudo-stem fibres, plantain bunch fibres and rice husk for construction application. Materials Circular Economy. 2021;**3**:15. DOI: 10.1007/s42824-021-00028-0

[19] Phung TA. Formulation et caractérisation d'un composite terre-fibres végétales : la bauge. [thesis]. France: Caen Normandie University; 2018

[20] Bui H, Sebaibi N, Boutouil M, Levacher D. Determination and review of physical and mechanical properties of raw and treated coconut fibers for their recycling in construction materials. Fibers. 2020;**8**(6):37. DOI: 10.3390/fib8060037

[21] Hussain M, Levacher D, Leblanc N, Zmamou H, Djeran-Maigre I, Razakamanantsoa A, et al. Properties of Mexican tropical palm oil flower and fruit fibers for their prospective use in eco-friendly construction material. Fibers. 2021;**9**:63. DOI: 10.3390/fib9110063

[22] Kesikidou F, Stefanidou M. Natural fiber-reinforced mortars. Journal of Building Engineering. 2019;**25**:100786. DOI: 10.1016/j.jobe.2019.100786

[23] Thi VVD. Matériaux composites à fibres naturelles / polymère biodégradables ou non [thesis]. Vietnam: Grenoble University, France-Ho Chi Minh University; 2011

[24] Hejazi SM, Sheikhzadeh M, Abtahi SM, Zadhoush A. A simple review of soil reinforcement by using natural and synthetic fibers. Construction and Building Materials. 2012;**30**:100-116. DOI: 10.1016/j.conbuildmat.2011.11.045

[25] Zmamou H. Eco-conception de nouveaux agromatériaux à partir de cendres de chaudière biomasse. Relation structures-propriétés. [thesis]. France: Rouen University; 2015

[26] Trujillo E, Moesen M, Osorio L, Van Vuure AW, Ivens J, Verpoest I. Bamboo fibres for reinforcement in composite materials: Strength Weibull analysis. Composites Part A: Applied Science and Manufacturing. 2014;**61**:115-125. DOI: 10.1016/j.compositesa.2014.02.003

[27] Fu SY, Lauke B, Mai YW. Science and Engineering of Short Fibre Reinforced Polymer Composites. 1st edit ed. London: CRC Press. eBook ISBN: 9781845696498

[28] Hine PJ, Duckett RA, Ward IM, Allan PS, Bevis MJ. A comparison of short glass fiber reinforced polypropylene plates made by

conventional injection molding and using shear controlled injection molding. Polymer Composites. 1996;**17**(3): 400-407. DOI: 10.1002/pc.10627

[29] Ferrara G, Coppola B, Di Maio L, Incarnato L. Tensile strength of flax fabrics to be used as reinforcement in cement-based composites: Experimental tests under different environmental exposures. Composites Part B: Engineering. 2019;**168**:511-523. DOI: 10.1016/j.compositesb.2019.03.062

[30] Prasad SV, Pavithran C, Rohatgi PK. Alkali treatment of coir fibres for coir-polyester composites. Journal of Materials Science. 1983;**18**:1443-1454. DOI: 10.1007/BF01111964

[31] Juenger MCG, Winnefeld F, Provis JL, Ideker JH. Advances in alternative cementitious binders. Cement and Concrete Research. 2011;**41**(12):1232-1243. DOI: 10.1016/j.cemconres.2010.11.012

[32] Bui H. Study on Performance Enhancement of Coconut Fibres Reinforced Cementitious Composites [thesis]. France: Caen Normandie University; 2021

[33] Paegle I, Fischer G. Evaluation of test methods used to characterize fiber reinforced cementitious composites. In: Proceedings of the International Conference Innovative Materials, Structures and Technologies. 2013. pp. 122-128

[34] Black L, Garbev K, Gee I. Surface carbonation of synthetic C-S-H samples: A comparison between fresh and aged C-S-H using X-ray photoelectron spectroscopy. Cement and Concrete Research. 2008;**38**(6):745-750. DOI: 10.1016/j.cemconres.2008.02.003

[35] Steiner S, Lothenbach B, Proske T, Borgschulte A, Winnefeld F. Effect of relative humidity on the carbonation rate of portlandite, calcium silicate hydrates and ettringite. Cement and Concrete Research. 2020;**135**:106116. DOI: 10.1016/j.cemconres.2020.106116

[36] Bui H, Boutouil M, Levacher D, Sebaibi N. Evaluation of the influence of accelerated carbonation on the microstructure and mechanical characteristics of coconut fibre-reinforced cementitious matrix. Journal of Building Engineering. 2021;**39**:102269. DOI: 10.1016/j.jobe.2021.102269

[37] Elsalamawy M, Mohamed AR, Kamal EM. The role of relative humidity and cement type on carbonation resistance of concrete. Alexandria Engineering Journal. 2019;**58**(4):1257-1264. DOI: 10.1016/j.aej.2019.10.008

[38] Yeon Ting MZ, Wong KS, Rahman ME, Meheron SJ. Deterioration of marine concrete exposed to wetting-drying action. Journal of Cleaner Production. 2020;**278**:123383. DOI: 10.1016/j.jclepro.2020.123383

[39] García-Maté M, De La Torre AG, León-Reina L, Aranda MAG, Santacruz I. Hydration studies of calcium sulfoaluminate cements blended with fly ash. Cement and Concrete Research. 2013;**54**:12-20. DOI: 10.1016/j.cemconres.2013.07.010

[40] AFNOR XP. P13-901. Compressed Earth Blocks for Walls and Partitions: Definitions – Specifications – Test methods. AFNOR editions, Paris, France; 2001. 35p

[41] Bauluz del Río G, Bárcena BP. Bases para el diseño y construcción con tapial. Madrid, Spain: Centro de Publicaciones, Ministerio de Obras Públicas y Transportes (MOPT); 1992

[42] Seifi S, Sebaibi N, Levacher D, Boutouil M. Mechanical performance

of a dry mortar without cement, based on paper fly ash and blast furnace slag. Journal of Building Engineering. 2019;**22**:113-121. DOI: 10.1016/j.jobe.2018.11.004

[43] ASTM. D790-03. Standard Test Methods for Flexural Properties of Unreinforced and Reinforced Plastics and Electrical Insulating Materials. West Conshohocken, PA: American Society for Testing and Materials; 2003. p. 9

[44] Cárdenas-Haro X, Todisco L, León J. Database with compression and bending tests on unbaked earth specimens and comparisons with international code provisions. Construction and Building Materials. 2021;**276**:122232. DOI: 10.1016/j.conbuildmat.2020.122232